Geology of
Offshore Ireland
and West Britain

Geology of Offshore Ireland and West Britain

D. NAYLOR

Northgate Exploration Ltd
Dublin

and

P. SHANNON

Department of Industry and Energy
Dublin

Figures prepared by
Mary Davies

Graham & Trotman

First published in 1982 by
Graham & Trotman Limited
Sterling House
66 Wilton Road
London SW1V 1DE

British Library Cataloguing in Publication Data

Naylor, D.
Geology of offshore Ireland and West Britain.
1. Ocean bottom 2. Atlantic Ocean
I. Title II. Shannon, P.
551.46'11 GC272

ISBN 0-86010-340-4 (hardback)
ISBN 0-86010-247-5 (soft cover)

Printed and bound in Great Britain by Robert Hartnoll Ltd, Bodmin, Cornwall
Phototypeset in Great Britain by Input Typesetting Ltd, London

Acknowledgements

The authors are conscious of their indebtedness to many people. We acknowledge the permission of the Department of Industry and Energy, Dublin, to publish some of the material. However, we wish to make clear that the views and opinions expressed herein are those of the authors, and are not necessarily shared by the Petroleum Affairs Division of the Department of Industry and Energy.

We express our thanks to Dr C. E. Williams, Director of the Geological Survey of Ireland, for encouragement and material support throughout the project. We are also grateful to Dr D. R. Whitbread, Senior Petroleum Advisor to the Department of Industry and Energy, Dublin, for advice and assistance.

Many people helped in preparation of the manuscript, but in particular we would like to thank Helen Cusack for much organizational work and aid with the references; also to Kathleen Malone and Margaret Donnellan for typing the manuscript with great accuracy and even greater tolerance. We are also in debt to the Cartographic Unit of the Geological Survey under the supervision of Amos Walsh, for the preparation of two of the text figures.

Clearly a work of this type rests on the labours of many academic and oil company scientists. The main references used in the preparation of each chapter are cited, but these listings are not comprehensive and we wish to apologize to the many geologists who have contributed important detail to the story presented here but who go unsung. Of the numerous colleagues who have given personal help and advice we are particularly grateful to Drs R. Scrutton and J. B. Megson for permission to use unpublished opinions, to Dr David Young for helpful discussion, to Nicola Barker on whose work in an earlier book we leaned quite heavily, and to Dr E. Dike for earlier collaboration in the Channel and Western Approaches areas. It goes without saying, however, that the authors accept full responsibility for any errors and omissions which are present in the book.

Finally we are both aware of how much we owe to our wives, Verney Naylor and Adrienne Shannon without whose support and understanding this project would never have started and certainly would never have arrived at a conclusion.

Contents

List of Figures

List of Tables

Chapter 1

Introduction

Historical Background

Only a few decades ago knowledge of the offshore geology west of Britain and Ireland was limited to a scattering of seabed samples. During the last twenty five years, however, the tempo of geological investigation has gradually increased. The framework and outline geology of the western seaboard is now known, although a great deal, some of it fundamental in nature, still remains to be elucidated. Several commercial and sub-commercial hydrocarbon discoveries have been made and are being produced, developed or investigated. A sustained phase of exploration drilling is anticipated in the next five years in the Western Approaches – Celtic Sea area and to the west of Ireland.

This book has grown out of an earlier small book[1] which appeared in 1975. Because of the volume of research and published data in the intervening years the earlier book was outdated to a degree that the face-lift of a second edition would not repair. In determining to write an entirely new book it was also decided to write the text at a slightly more technical level. We felt that the avoidance of technical terms attempted in the earlier book detracted from its use by geologists, without necessarily making it significantly easier for the committed non-technical reader. This volume is aimed towards the undergraduate geology student, but we hope that it will also be of use to the professional geologist seeking background material. Appendix 1 and the Glossary (Appendix 2) contain introductory material regarding geology and petroleum exploration techniques which we trust will help the non-specialist, in conjunction with some of the more general references cited in the text, to follow the main parts of the chapters and the text figures. This first chapter also contains a small amount of introductory material concerning the opening of the North Atlantic Ocean which should help to lead into the more detailed account which follows, and also help in the understanding of some of the points made in later chapters.

Credit must go to university departments, and later the Institute of Geological Sciences, for their parts in initially establishing the existence of, and geological outlines for, the offshore basins along the western seaboard of Britain and Ireland. This was achieved by programmes of shallow seabed sampling, by detailed gravity and magnetic measurements, and by shallow seismic profiling, across the continental shelf. In the main this was carried out over two decades beginning in the mid 1950s. Some areas, notably the Channel and Western Approaches, had seen even earlier pioneering work by the universities. The teams from these institutes established the presence of thick Mesozoic-Tertiary basins on the continental shelves around Britain and Ireland. From the geophysical measurements it was also possible to arrive at some estimate of the thickness of the Mesozoic-Tertiary sediments infilling these basins.

Exploration of the western offshore basins by the oil industry has built upon the base provided by the academic institutions. The basins have been traversed by many tens of thousands of miles of reflection seismic profiling. The reader is referred to Appendices 1 and 2 which cover some of the basic oil company exploration and drilling techniques. Interpretation of the seismic data was followed by drilling in most of the basins shown on Fig. 1.2. Knowledge of the subsurface geology and structure within the basins has therefore greatly increased and in turn allows a better understanding of the development of the western seaboard of Europe and of the palaeogeography at different times in the past.

Relatively few years ago the study of the geology of the continental shelf west of Ireland and Britain was regarded as an academic pursuit, for which relatively few funds were made available. That there has been a dramatic change in this picture is due to two main factors. The first was the discovery of oil and gas in the North Sea which generated interest in other sedimentary basins on the shelves around these islands and improved deep water technology. The second factor has been the cost of energy which has made search for oil economic in more hostile environments. Governments have also become more aware of the need for indigenous energy supplies.

The involvement of the petroleum industry in the Northwest European continental shelf spans a little more than twenty years. In 1959 Shell/Esso discovered the giant Groningen Gasfield in northeastern Holland, one of the largest gasfields in the world. The focus then moved to the younger rocks which were believed to cover the Southern North Sea area, linking the sedimentary basins of England with those of Holland and north Germany. At first the main target in the Southern North Sea was the Permian sandstone gas reservoir, productive in the Groningen Field. However as more insight was gained into the overall structure and geology of the North Sea Basin the search spread northwards and also examined other potential hydrocarbon-bearing horizons.

The huge success of the search for oil in the Northern North Sea is now known to everyone. This success has guaranteed a continued interest by the industry in the continental shelf lying west of Britain and Ireland. The generally unpromising appearance of the Precambrian and Palaeozoic metamorphosed rocks which form much of Ireland and the western part of England and Scotland was an early factor detracting from interest in the western shelf area. However it was gradually realized that the small patches of younger Mesozoic and Tertiary sediments which locally overlie the metamorphosed rocks, particularly in the coastal belt, were in fact the onshore margins of sedimentary basins stretching westwards onto the continental shelf. In some cases, as in the west of Ireland, there is no onshore remmant and basins have been located entirely by exploration of the shelf area.

The Continental Margin

The morphology of the continental shelf west of Britain and Ireland is shown on Fig. 1.1, and further detail is shown on text figures throughout the book. The continental margin is wide by general world standards, being greater than 200 nautical miles in many places, particularly in the region of the Rockall Plateau. In the south, off the Western Approaches, the general course of the continental edge is curved and linear, the continental slope being cut by deep canyons. The continental shelf in this region slopes gradually westwards from the land and is generally devoid of prominent topographic features except along the trend of the westward continuation of the Cornubian Peninsula where the Scilly Isles and various subsea provinences mark the site of granite intrusions.

North of this ridge, however, the continental margin is morphologically more complex. The foot of the continental slope swings out westwards from Ireland around the Rockall Plateau. The Porcupine Seabight and Rockall Trough are deep water embayments east of this line and separate a number of higher plateau areas on the shelf. Even the plateaux, most notably the Rockall Plateau, have varied topography and occasionally become very shallow. It will be seen in later chapters that there is strong geological control upon the morphology of the western shelf areas and that the deeper water areas are generally underlain by thicker developments of the Mesozoic and Tertiary sedimentary rocks.

The general outlines of the thick post-Carboniferous basins discovered by academic and oil industry activity over the last few decades are shown in Fig. 1.2. The basins are varied both in the thickness and nature of the Mesozoic and Tertiary sediments which comprise the fill. However, the formation and opening of the North Atlantic which is described later in this chapter has exercised a strong overall control on basin development. As would be expected in an area subjected to a persistent

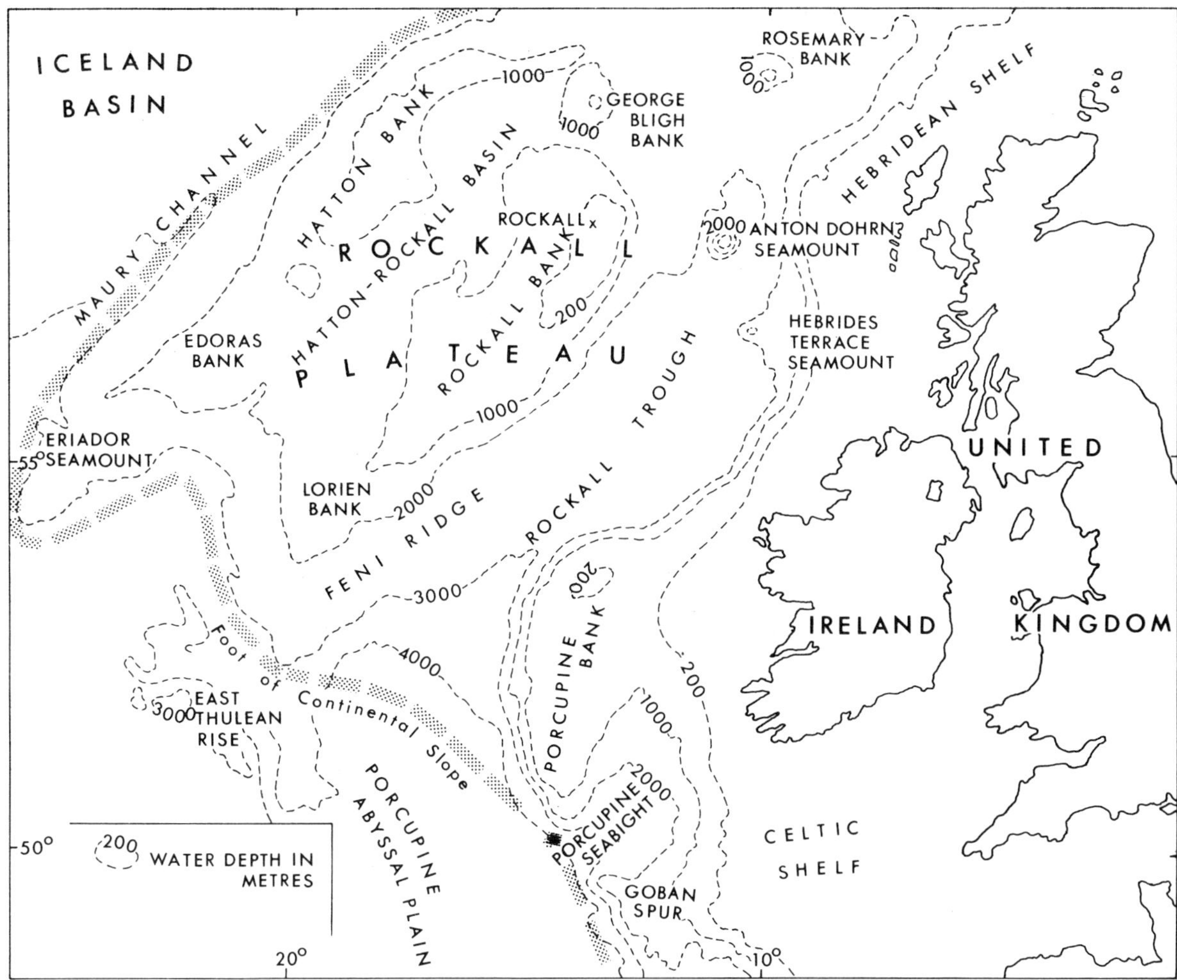

Figure 1.1. Morphology of the Continental Shelf west of Britain and Ireland. (Depth contours after GEBCO Map 5.04 Scale 1 : 10 million, 1978).

tectonic tensional regime, major faults control the basin margins. The pre-existing structural framework also exercised strong control over the trend of the basins. This is particularly evident in the fringe of basins extending along the western coastline of Scotland south from the Shetland Islands. Here the Caledonian grain shows in the trend of the younger basins, and the older faults have been rejuvenated along the basin margins. The interaction of the deeper Caledonian pattern and younger fault systems is more complex in the Irish Sea and Celtic Sea areas but some degree of Caledonian imprint is generally present. The basins west of Ireland cut obliquely across the earlier structural trends, but even here earlier major fault lines have had considerable influence on basin development.

The westwards extension of major structural lineaments of the mainland, such as the Great Glen and Highland Boundary Faults, and their correlation with continental shelf features has been the subject of many academic papers. There is no general agreement at the present time regarding the precise westward path of these lineaments.[11]

The precise limits of British and Irish offshore sovereignty on the continental margin remain to be determined. The lines dividing the Continental Shelf as between Britain and Ireland in the Irish and Celtic Seas, and in the northwest, have not been agreed. The two parties have agreed to go to Arbitration on the issue and the problem should be resolved in the next few years. Similarly the precise position of the outer limit of the offshore sover-

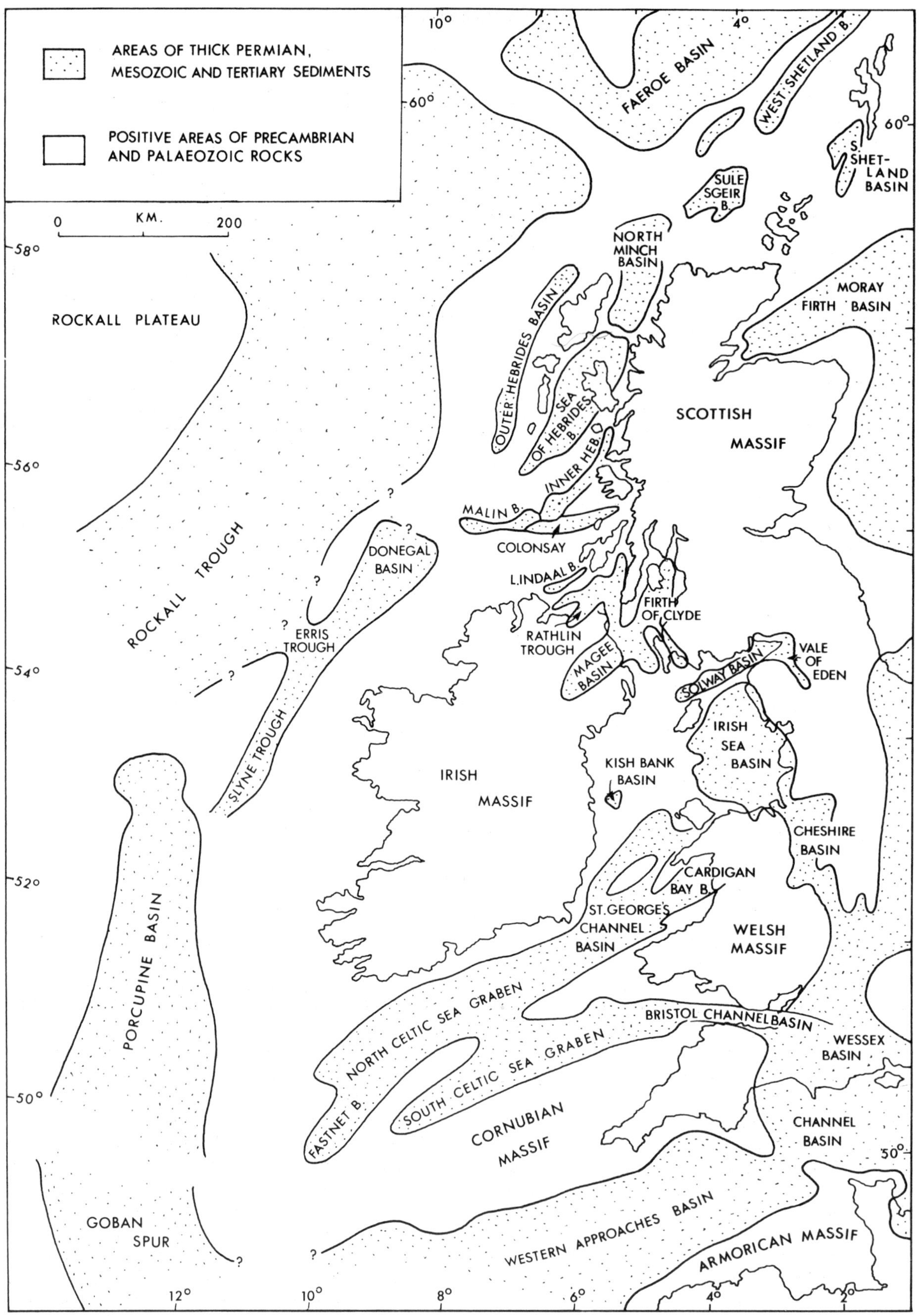

Figure 1.2. Sedimentary basins west of Britain and Ireland.

eignty of both countries is not known. The Geneva Convention of 1958, which Britain but not Ireland ratified, allowed a nation to exercise sovereignty out to a water depth of 200 m and beyond this to a point at which water depths imposed a technical limit to exploration. Many nations, including the United Kingdom and Ireland, have designated deeper waters than 200m and in some cases granted licences in them. The United Nations Law of the Sea Conference is considering this problem, along with many others. When finally ratified the new Convention which is being prepared will include formulae for the definition of the outer limit of the Continental Shelf. In the meantime however the issue is unresolved. Technology and the demand for petroleum have pushed national offshore boundaries out into progressively deeper water, and this has been seen west of Britain and Ireland as elsewhere.

Opening of the North Atlantic

The concepts of *plate tectonics* and *sea-floor spreading* were developed in the 1960s, more or less in line with the increasing knowledge of the offshore basins on both sides of the North Atlantic ocean. These concepts have been a great aid to the understanding of the processes and history of the offshore basins.

The entire surface of the earth is composed of a number of internally rigid but relatively thin (95–160 km thick) interlocking plates continuously but very slowly in motion relative to one another. The upper part of each plate is composed of either thick (24–40 km) low density continental crust or of thinner (8 km) denser oceanic crust. Continental crust characterizes the land regions of the world and their adjacent continental shelves and oceanic crust characterizes the floors of the major oceans. The lower zone of the plate extends beneath oceanic and continental crust alike and is composed of the uppermost section of the high density mantle. The relatively rigid plates move on the more mobile underlying mantle layers which constitute a major portion of the earth's interior. Where the plates move towards each other plate material is consumed at a destructive boundary as one plate is thrust down beneath the other into the mantle. On other margins however, new ocean floor is formed by volcanic outpouring as the plates move apart. It is this latter configuration which is of interest to us in a consideration of the North Atlantic.[1–4]

Prior to the initiation of ocean-floor spreading the North Atlantic continents formed a single continental plate in which the present day North Atlantic Ocean, including the Bay of Biscay, did not exist. The reconstruction shown on Fig. 1.3 depicts the plates prior to separation except that Spain should be rotated northwards to close the Bay of Biscay as shown by the arrow. It is clear from this configuration that the pre-spreading geological history of what is now the East Canadian shelf and of the nearby shelf west of Britain and Ireland must have been very similar. The Caledonian mountains extended as a single chain from Norway across Britain and Ireland and down the eastern zone of Canada and the United States. Then, in Mesozoic times, grabenal troughs developed in response to the tensional stresses which were eventually to rupture the plate and form the new Atlantic Ocean. It was during this phase that the framework of the basins which are described in this book was established.

Reconstructions of the type shown in Fig. 1.3 are achieved mainly by consideration of the trend of magnetic stripes on the floor of the ocean. The mid-ocean ridge which constitutes the active spreading axis between the Eurasian and North American plates is formed of molten material forced to the surface. As the material cools it is magnetized in the direction of the earth's field at that time. This magnetized material is eventually pushed away to each side of the ridge as more material is extruded. As a result of periodic reversals of the earth's magnetic field a series of equivalent and symmetrical magnetic stripes have been formed on each side of the Mid-Atlantic Ridge with the widening of the ocean (Fig. 1.4). These magnetic lineations have a strong linearity over distances of hundreds of kilometres, trending parallel to the margins of the ocean basin.[6] Each magnetic stripe has been designated by number, starting with the most recent example. The reconstruction in Fig. 1.3 has been produced by narrowing the North Atlantic back to the earliest of these magnetic anomalies. By superimposing a pair of these lineations of a particular age, the original width of the oceanic crust separating the continents of that period becomes apparent.

The reader wishing to cover more of this subject before proceeding further is referred to a number of basic texts on seafloor spreading and the history of the North Atlantic.[1–6]

TABLE 1.1 General Statigraphic Classification

Era	*System (Period)*	*Series (Epoch)*	*Stage*	*Age My B.P.*
CENOZOIC	QUATERNARY	Holocene		
		Pleistocene		
			Sicillian	
			Calabrian	1.8
	TERTIARY	Pliocene	Plaisancian	
			Zanclian	5.0
		Miocene	Messinian	
			Tortonian	
			Helvetian	
			Burdigalian	
			Aquitanian	22.5
		Oligocene	Chattian	
			Rupelian	37.5
		Eocene	Priabonian (Bartonian)	
			Lutetian	
			Ypresian	53.5
		Palaeocene	Landenian (Thanetian)	
			Montian	
			Danian	65
MESOZOIC	CRETACEOUS	Upper (Late)	Maastrichtian	
			Campanian	
			Santonian	
			Coniacian	
			Turonian	
			Cenomanian	100
		Lower (Early)	Albian	
			Aptian	
			Barremian	
			Hauterivian	
			Valanginian	
			Berriasian	140
	JURASSIC	Upper (Late)	(Purbeckian) Portlandian	
			Kimmeridgian	
			Oxfordian	
		Middle (Middle)	Callovian	
			Bathonian	
			Bajocian	
		Lower (Early)	Aalenian	
			Toarcian	
			Pliensbachian	
			Sinemurian	
			Hettangian	195
	TRIASSIC	Upper (Late)	Rhaetian	
			Norian	
			Carnian	
		Middle (Middle)	Ladinian	
			Anisian (Virglorian)	
		Lower (Early)	Skythian (Werfenian)	225
PALAEOZOIC	PERMIAN			280
	CARBONIFEROUS			345
	DEVONIAN			395
	SILURIAN			435
	ORDOVICIAN			500
	CAMBRIAN			570
PRECAMBRIAN				

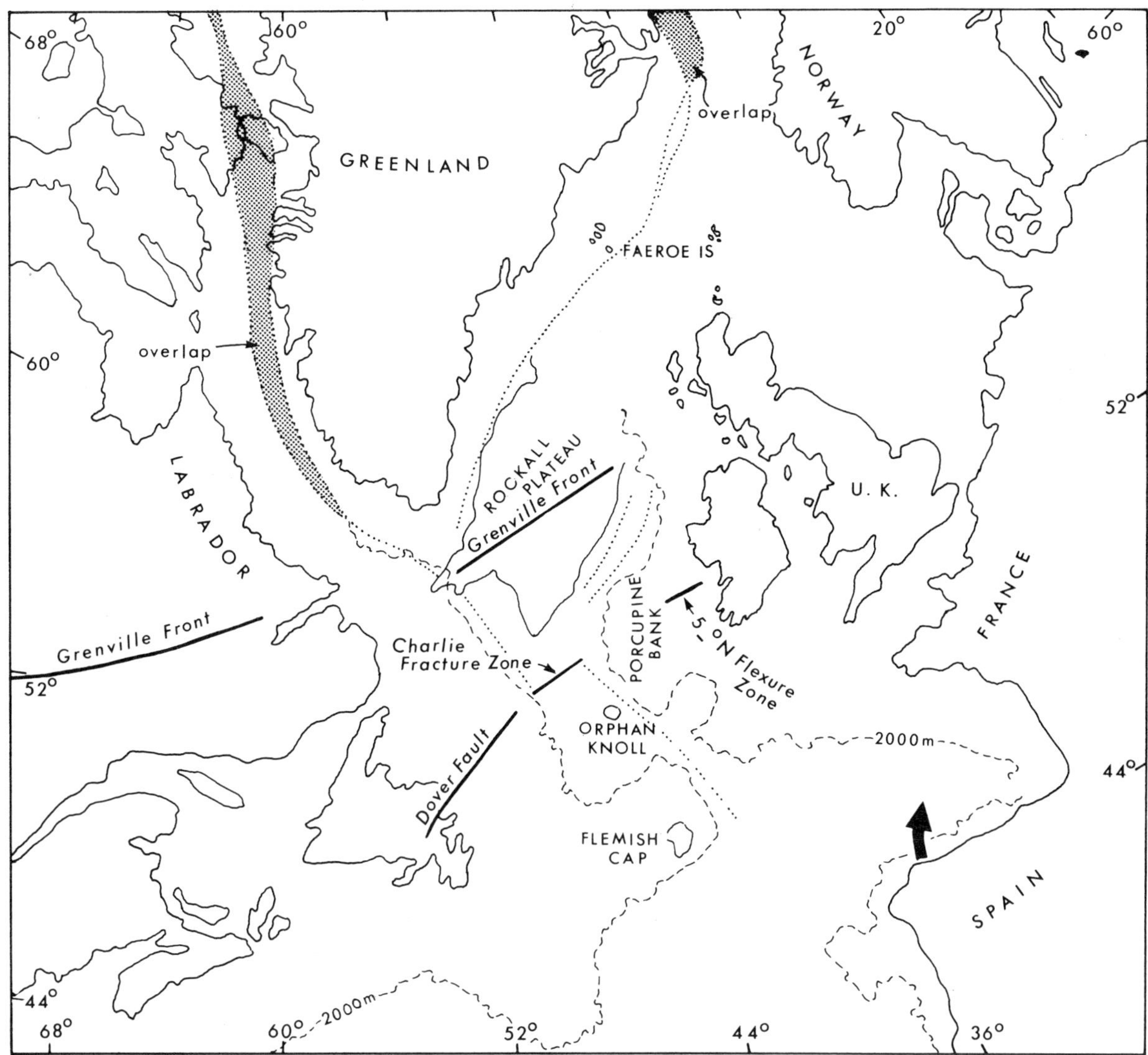

Figure 1.3. Reconstruction of the palaeogeographic positions of Greenland and Europe relative to North America at the time of initial opening of the North Atlantic (redrawn from Srivastava (1978)).[9] Note that Iberia should be rotated towards France to give a correct position, as shown by the black arrow.

From a study of the pattern of magnetic stripes in the North Atlantic, in conjunction with the results of the Joides (Joint Oceanographic Institutions Deep Earth Sampling) deep sea drilling programme, it can be established that prior to 200 million years ago (Late Triassic time) the North Atlantic did not exist as an ocean, and that the continental plates of North America, Greenland and Western Europe were joined together to form one huge landmass (Fig. 1.3). Sometime during the Middle Jurassic period an embryonic plate margin began to develop between the North American continent to the west and Europe to the east.

During the subsequent period from mid-Jurassic to mid-Cretaceous times a continuous, but slow separation of the continental blocks of southern Europe/North Africa and North America led to new oceanic crust being generated over the entire length of the southern North Atlantic. To the north the development of the plate boundary between the potential Greenland and North American Plates to the west and the European Plate to the east had only reached a preliminary stage during this time period, with tensional stresses in the crust initiating the chain of asymmetric basins and fault-bounded troughs along the west coast

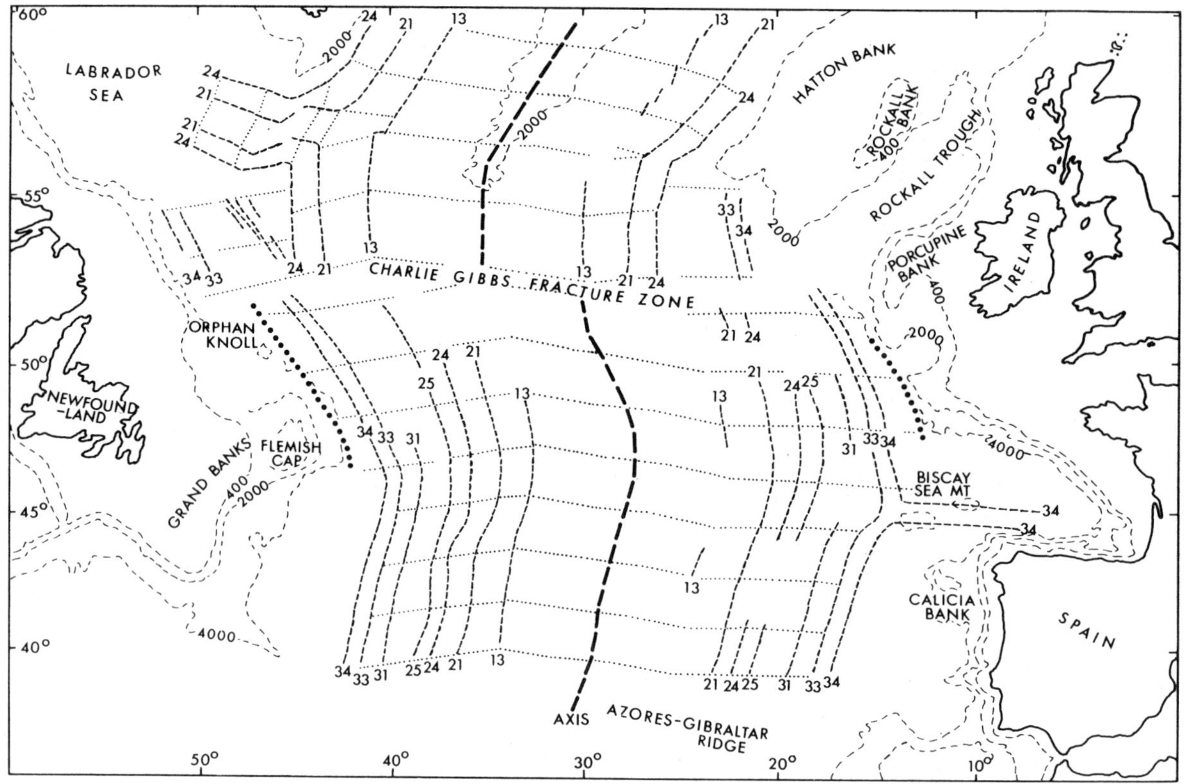

Figure 1.4. Present magnetic anomaly trends in the North Atlantic showing the various pairs of positive anomalies on each side of the present axis of spreading (Reykjanes Ridge). Modified after Talwani (1978).[10] Presumed boundary between continental and oceanic crust east of Flemish Cap and west of Ireland is indicated by a coarse dotted line. Fine dotted lines are flowlines indicating the relative motion between North America and Europe. Bathymetric contours are in metres). The anomaly numbers on this diagram, e.g. anomaly 34 differ from those used by other authors (for discussion see Chapter 9).

of Britain and Ireland. Thinning of continental crust probably took place during this period along zones of tension. One of the problems in diciphering the early history of sea-floor spreading in the North Atlantic is the lack of magnetic stripes in the magnetic quiet zone corresponding to 80–110 My (Cretaceous).

We are accepting here the ages shown in Table 1.2 for some of the key lineations.[9] By 115 My the Grand Banks and Iberia were separated. Iberia then rotated away from France, thus producing the Bay of Biscay and this spreading arm continued to operate until about 75 My ago. However, Aptian – Albian sediments known in the Bay of Biscay do not continue onto ocean crust west of Goban Spur, which is therefore post-Albian (100 My) in age.[12] South of the Charlie Gibbs Fracture Zone, in the Porcupine Bank area, there is evidence (see Chapter 9) that spreading has initiated before anomaly 32 (75 My) time. Some workers have identified older anomalies, e.g. anomaly 34 (90 My[9]) denoting the onset of spreading but the precise history of the Porcupine – Rockall area is still uncertain. A zone of extension along Rockall Trough resulted in thinning of the continental crust but it is uncertain whether new oceanic crust was generated on the floor of the Trough (for discussion see Chapter 9). Upper Mesozoic rifting certainly failed to extend northward along this line to separate Norway from Greenland.

Table 1.2 Ages of some of the key lineations

Anomaly No.	*Age My*		*Epoch*
7	27	Oligocene	
13	40	Eocene	
			Tertiary
21	53	Eocene	
24	60	Paleocene	
32	75	Campanian	Late
(34)	(90)	Campanian	Cretaceous

Certainly, spreading was underway between Ireland (Rockall-Porcupine) and Canada (Newfoundland – Labrador) by 75 My (anomaly 32) ago. This spreading axis extended at the same time into the southern Labrador Sea between Greenland and Canada.[9] Spreading extended into the northern Labrador Sea during the Maastrichtian (anomaly 28). Crustal stretching and block faulting took place at this time in the grabenal basins off eastern Canada. The Labrador Sea spreading axis continued until about 40 My.

The opening of the northern North Atlantic between Greenland and Northwest Europe was not firmly established until Palaeocene times, although earlier but unsuccessful attempts to extend the North Atlantic oceanic spreading system north of latitude 54° indicate the presence of a long-existing zone of crustal weakness between the two continental plates. Rifting between the Greenland and Eurasian plates may have started as early as Campanian (Upper Cretaceous) time (anomaly 32:75 My) and thus represent an embryonic stage in the break-up of the Northwest European – Greenland continental block which failed to become sufficiently advanced to have formed a significant zone of new oceanic crust, if any.

At anomaly 24 time (60 My) in the Palaeocene, sea-floor spreading began between Ireland and Greenland and northwards in the Norwegian Sea. This new spreading arm intersected the existing axis of spreading to form a triple junction at a point between the tip of Greenland and Rockall Plateau. The southern (Ireland – Labrador) and northwestern (Southern Labrador Sea) arms suffered readjustment of alignments at this time. Until anomaly 13 (40 My) this triple junction continued to operate, although spreading rates in the Labrador Sea slowed considerably. The result, therefore, was that Greenland was progressively separated from the North American and European Plates.

The Labrador Sea spreading arm failed some 40 My ago and since that time the remaining two arms have formed the Mid-North Atlantic Ridge. Spreading has continued through the Tertiary to produce the present anomaly picture shown in Fig. 1.4. Throughout the Tertiary there has been a considerable variation in the spreading rate across the individual portions of the ridge giving rise to symmetrical bands of oceanic crust on either side of the axis, each characterized by a distinctive magnetic and morphological pattern. Important changes of spreading rate occurred between 35 and 45 My and between 10 and 15 My ago, which affected sedimentation on the adjoining margins.

The present day median position of the Reykjanes Ridge between the continental shelves of Southeast Greenland and Rockall Plateau suggests that spreading has always proceeded from a single northeast trending axis.

North of Iceland the southern Norwegian Sea ridge also continued active spreading through Early Tertiary times. In the Early Oligocene the new spreading ridge was closely parallel to the Greenland continental shelf along most of its length, but, just to the north off Scoresby Sound, Greenland, the ridge moved westwards at about anomaly 13 time, breaking off a small fragment of continental crust, the Jan Mayen microplate. This fragment now lies approximately 350 km away from the Greenland shelf edge, entirely surrounded by oceanic crust.

This brief account of the opening of the North Atlantic is intended to set the scene for the following chapters. The form of the basins and the early sedimentary fill was directly related to the early tensional history caused by seafloor spreading. The later sedimentary history on both sides of the new ocean was strongly affected by the pulses of spreading movement.

References

1. NAYLOR, D. and MOUNTENEY, S.N. 1975. *Geology of the North-West European Continental Shelf Vol. 1.* Graham, Trotman & Dudley Ltd. London, 162 pp.
2. TARLING, D. H. and TARLING, M. P. 1971. *Continental Drift.* G. Bell & Sons Ltd. London, 112 pp.
3. LE PICHON, X. 1968. Sea-floor spreading and continental drift. *J. geophys. Res.* **73**.
4. DEWEY, J. F. and BIRD, J. M. 1970. Mountain belts and the new global tectonics. *J. geophys. Res.* **75**.
5. LE PICHON, X., SIBUET, J. C. and FRANCHETEAU, J. 1977. The fit of the continents around the north Atlantic Ocean. *Tectonophysics,* **38**, 169–209.
6. LAUGHTON, A. S. 1975. Tectonic evolution of the Northeast Atlantic Ocean: a review. *Nor. Geol. Unders. Publ.* **316**, 169–193.
7. LAUGHTON, A. S. 1972. The South Labrador

Sea – a key to the Mesozoic and early Tertiary evolution of the North Atlantic. Initial Reports of the Deep Sea Drilling Project, 12, USPGO. Washington, 1155–1179.

8. HAWORTH, R. T. 1980. Appalachian structural trends northeast of Newfoundland and their trans-Atlantic correlation. *Tectonophysics* **64**, 111–130.
9. SRIVASTAVA, S. P. 1978. Evolution of the Labrador Sea and its bearing on the early evolution of the North Atlantic. *Geophys. J. R. astr. Soc.* 313–357.
10. TALWANI, M. 1978. Distribution of basement under the eastern North Atlantic Ocean and the Norwegian Sea: *in* Bowes, D. R. and Leake, B. E. (*Eds*). Crustal evolution in northwestern Britain and adjacent regions. *Geological Journal, Special Issue No. 10*, 348–375.
11. BAILEY, R. J. 1979. The continental margin from 50°N to 57°N: its geology and development: *in* Banner, F. T., Collins, M. B. and Massie, K. S. (*Eds*). *The North-West European Shelf Sea 1. Geology and Sedimentation.* Elsevier, Amsterdam, 11–24.
12. ROBERTS, D. G., MASSON, D. G., MONTADERT, L. and DE CHAPELO, 1981. Continental Margin from the Porcupine Seabight to the Armorican Marginal Basin: *in* Illing, L. V. and Hobson, G. D. (*Eds*). *Petroleum Geology of The Continental Shelf of Northwest Europe.* Heyden & Son Ltd, London, 455–473.

Chapter 2

Channel Basin

Introduction

The Channel Basin is a sedimentary trough lying between the south coast of England and Brittany, beneath the eastern portion of the English Channel. For the purposes of this discussion its eastern boundary will be taken at about 1°E longitude. The western boundary, against the Western Approaches Basin, is taken at an important positive axis of basement rocks which runs obliquely across the Channel from Start Point to Cherburg; the *Start – Cherbourg* (or Contentin) *Ridge*. With the exception of the eastern extension of the Hurd Deep water depths are everywhere less than 100m, and over considerable areas of the eastern Channel less than 50m.

In some respects the Channel Basin may seem a strange starting point for a discussion of the offshore basins on the west side of Britain and Ireland. However, there is good reason for beginning here and working outwards. As we have seen in the introductory chapter, a fringe of elongate fault-controlled Mesozoic-Tertiary basins rims the western seaboards of Britain and Ireland, which are themselves composed in the main of older rocks. In the case of some of the basins, for example along the Scottish coastline, the basin margins are exposed on the adjacent coastline. The basins around the southern part of Ireland, on the other hand, are entirely offshore and do not extend onto the Palaeozoic mainland massif of Ireland. The presence of these basins was unsuspected until two decades ago. The Channel Basin is unique in this context in that it extends onshore into southern England and northern France where outcrops and onshore wells allow a full examination of the Mesozoic stratigraphy. The classic sections along the Devon, Dorset and Hampshire coastline of England are particularly useful in this regard. In terms of published data the level of knowledge concerning the stratigraphy of the Channel Basin is not matched further westward, until we arrive at the offshore basins of eastern Canada. For this reason an outline account will be given of the onshore stratigraphy.

It can be anticipated that the stratigraphic sequence exposed along the south coast of England and the Isle of Wight will be similar to that of the eastern part of the English Channel. This degree of correlation will naturally decrease westwards away from the onshore sections into the Western Approaches and Celtic Sea Basins and beyond. Knowledge of the geological sequence offshore over the British sector of the Channel is drawn largely from the considerable amount of seismic reflection and seafloor sampling data obtained over many years[1–3] by academic institutions, together with research cruises undertaken by the Institute of Geological Sciences. French scientists have also been active in gathering geological and geophysical data over the offshore portions of the basin.

The offshore outcrop pattern of the Channel Basin is controlled by a number of major folds and faults (Figs. 2.1 and 2.2). In structural terms two provinces may be recognized. What

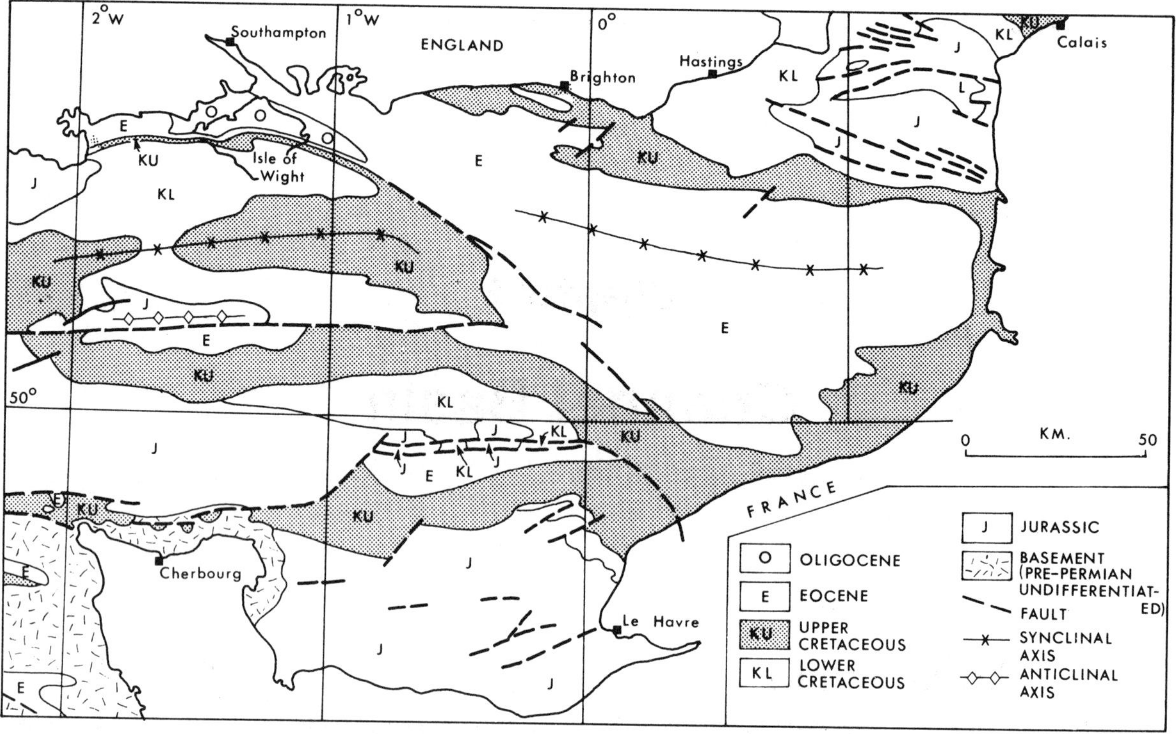

Figure 2.1. Sketch map of the solid geology of the English Channel (simplified after B.R. G.M. (1974)[5] 'Geological Map of the English Channel', with slight modifications).

may be termed the *Central Channel* lies in the area between the Start-Cherbourg Ridge and the southeast trending Bembridge-St. Valery structural line which links the Purbeck-Isle of Wight Disturbance in England with the Pays de Bray Fault in France. The dominant structural grain in the Central Channel is east–west and the area is dominated by three monoclines which involve Jurassic, Cretaceous and Eocene strata. These structures are the North Cherbourg, Mid-Channel (South Wight) and Purbeck-Wight structures. They probably overlie deep basement fracture zones which have undergone repeated phases of movement. The intervening areas are disposed in more open folds.

The *Eastern Channel* is dominated by a gently dipping syncline, which comprises in part the Tertiary Dieppe Basin resting on a subsided Basement Platform. Structural trends in this region are northwest-southeast and the Bembridge – St. Valery line is a continuation of the Birr structure onshore with this trend. There is a distinct swing in the structural grain from northwest-southeast in the Eastern Channel, to east-west in the Central Channel and west southwest in the Western Approaches.

Geophysical Data

A systematic collection of geophysical data and some 280 sea-floor rock samples over the eastern part of the English Channel was made between 1968 and 1971[3] and forms an easterly continuation of an extensive work programme carried out earlier over the Western Channel and Western Approaches.

The geophysical work yielded approximately 3,500 nautical miles of shallow seismic reflection, magnetic and echo sounding data. Coring of the sea bed on a 2.5 nautical mile grid was also carried out. In order to delineate the major geological structure of the basin, the grid traverses were run at right angles to the anticipated trend of the folds. Although the maximum depth of seismic reflection information did not exceed 300m, the penetration was sufficient to be able to pick out the major structural features. In addition, the magnetic data gave further information on the variation in the depth to the underlying Precambrian and

Palaeozoic basement surface.

The Institute of Geological Sciences has also undertaken a number of cruises in the Western Approaches and extending eastwards into the Channel Basin, using the drillships m.v. Whitehorn and m.v. Sealab to make shallow boreholes in addition to gravity and vibrocoring sample collection.[4] Compilations of these data have appeared in a number of reports and on 1:250,000 scale geological maps sheets.

The Channel south of latitude 50° North has been the subject of intense French research work. A coordinated programme of bottom sampling and shallow seismic profiling was carried out by the Bureau of Geological and Mineral Research, the French Geotechnical Society and the Service Géologique National. Three 1:250,000 maps covering the central and eastern parts of the Channel and incorporating the results of this work were published in 1971 by the Service Géologique National, and these were followed in 1974 by a 1:1,00,000 compilation Geological Map of the English Channel.[5] The results of British and French university and institute research were well summarized in the proceedings of a conference published by the Royal Society in 1975.[6,7]

Deep seismic reflection programmes have been carried out for the oil industry since that time but few of the results have been published. The first oil company deep drilling has only recently taken place so that interpretation of the basin sequence relies heavily on the published seabed sampling results.

Onshore Stratigraphy

Pre-Permian rocks are exposed in Brittany and in the Cornubian Peninsula and have also been encountered in deep onshore boreholes. At Start Point (Devon) and Lizard Point (Cornwall) strongly metamorphosed rocks, including serpentinites, are in contact with folded Devonian strata. The thick marine Devonian of the Cornubian Peninsula comprises slates with interbedded limestones, sandstones, and vol-

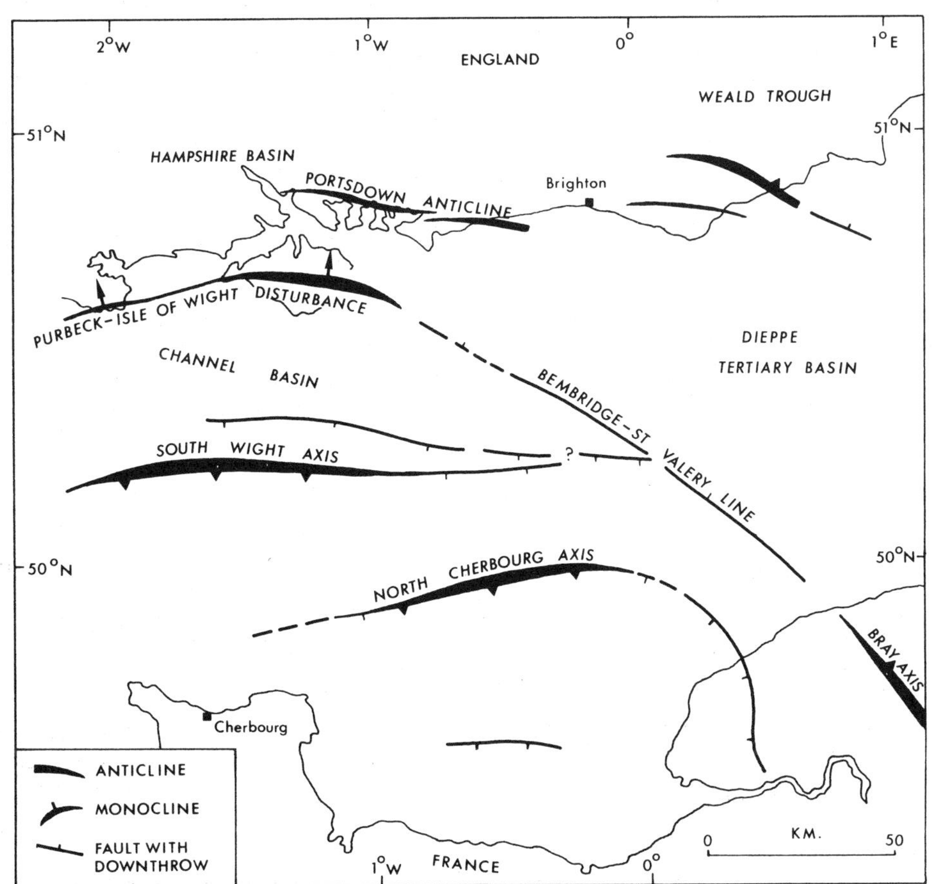

Figure 2.2. Major structural elements of the Channel Basin (modified after Smith and Curry (1975)[6]).

canic horizons. Carboniferous rocks crop out mainly in North Devon where the Upper Carboniferous 'Culm' facies consists of folded and cleaved slates and turbiditic sandstones.

The Palaeozoic rocks of Brittany comprise low-grade metamorphic shales, schists, and quartzites of Lower Palaeozoic age including Cambrian phyllites and psammites, and Ordovician quartzites. On the Contentin Peninsula sericitic phyllites and schists of Precambrian age are exposed. The Palaeozoic rocks are highly tectonized and folded into the Precambrian basement which is intruded by massive granite batholiths.

A number of deep wells along the coastal belt of southern England and on the London Platform have reached the Palaeozoic floor (Figs 2.3 and 2.4). Wells on the London Platform bottomed in Devonian mudrocks and the Brightling-1 well in Kent reached total depth in grey and brown Lower Devonian rocks. The Bolney-1 well, drilled by Esso, passed from the Jurassic through Carboniferous limestones into Middle Devonian strata. Similar rocks, possible Lower Devonian, were encountered at the base of the Ashdown-1 well. Further east, phyllites of probable Middle or Late Devonian age were penetrated by one well in the Wytch Farm Oilfield in Dorset. Lower Carboniferous rocks are believed to have been encountered in deep exploratory wells on the coast of Sussex, but Namurian strata have not been proven either along the south coast or in the Weald area in general. In Kent, Coal Measures rest with angular unconformity on Lower Carboniferous rocks and overstep northwards and eastwards onto older rocks. More than 800m of Coal Measure strata are present, with equivalents of the main threefold division of the Coal Measures elsewhere in Britain. The succession is divided into an upper Sandstone Division and a lower coal-bearing Shale Division.

It seems from available borehole evidence therefore that the London Platform is comprised at depth of Devonian and older Palaeozoic rocks. Basement beneath the Weald and Hampshire Basins appears in the main to consist of folded and faulted Devonian and Carboniferous rocks. Although the trend of folding in the Kent Coalfield is charnoid the basement grain further west appears to be more nearly east-west (Variscan). Interbedded red shales, carbonates and sandstones in the Pas de Calais region of France, are of Middle-Upper Devonian age and rest unconformably on folded Lower Palaeozoic shales.

Permo-Triassic rocks are not present in Kent or in the wells of the Weald. Westwards a wedge of redbeds is developed between the Jurassic and the Palaeozoic basement. Excellent coastal exposures extend from Dorset to Devon. The sequence is unfossiliferous and is divided into Permian and Triassic purely on lithological grounds. Suggested correlations and recommendations for lithostratigraphic nomenclature have been made by Warrington and his fellow workers.[9] Along the south Devon coast the total sequence may exceed 2,750m in thickness. The lower part of the succession (Permian) comprises red breccias, sandstones and siltstones resting unconformably on folded older rocks. There was considerable relief within the depositional basin at this time and high mountains (more than 3,500m) in the Cornubian hinterland. The eastern limit of the presumed Permian in the subsurface is not known, although the Nettlecombe-1 well near

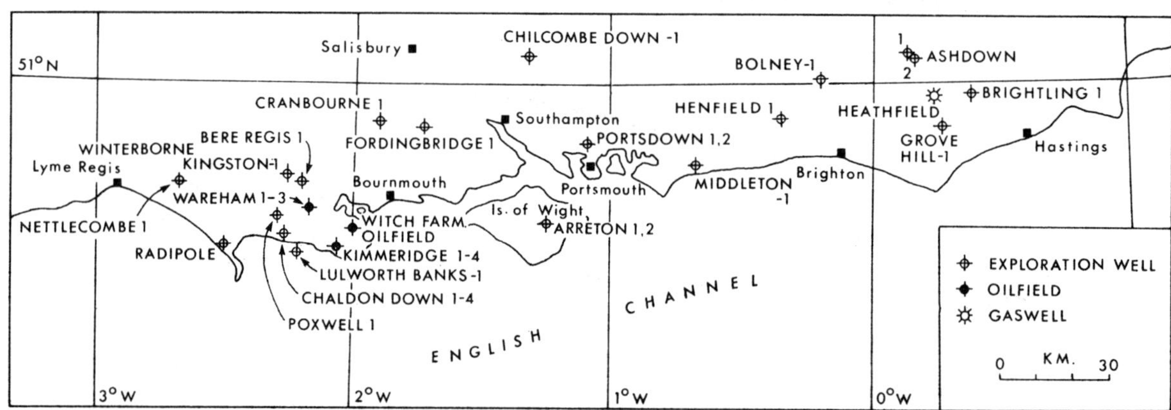

Figure 2.3. The locations of some deep exploratory boreholes in southern England.

Bridport, Dorset (Fig. 2.3) is rumoured to have penetrated a thick Permian and Triassic section before entering Lower Carboniferous shales. A borehole at Puriton in Somerset also penetrated some 175m into Permian marls and sandstones. Almost 1,000m of redbeds are ascribed to the Permian in wells at the Wytch Farm Oilfield Dorset.[11] It seems likely therefore that the 'Permian' extends at least as far east as Southampton Water.

The Triassic sequence consists of a lower pebble bed and alluvial sandstone section (Sherwood Sandstone Group). Sandstones within this unit provide the lower producing reservoir in the Wytch Farm Oilfield in Dorset. These coarse beds grade upwards to the Upper ('Keuper') Marls (Mercia Mudstone Group), comprising red and green mudrocks and marls with sandstone, carbonate and evaporite intercalations. A coal boring at Lyme Regis, Dorset, in 1901 proved a thickness approaching 350m of the Upper Marls. There is a considerable development of halite in the Keuper of the Winterborne Kingston borehole (Fig. 2.2).[12] The eastward extension of this unit is not precisely known but the basal parts of the Portsdown (Fig. 2.4) and Kingsclere wells just penetrated redbeds with evaporites. However in the Henfield boring the Jurassic rested unconformably on Carboniferous rocks. The Triassic basin therefore probably extended further east than the area of Permian deposition.

Wells in the Wytch Farm Oilfield have penetrated the full Permo-Triassic section and one reached phyllites of probable Middle-Late Devonian age at its base. There is an angular unconformity within the Permo-Triassic sequence and this is taken[11] as the base of the Triassic, although there is no palaeontological evidence. Geologists working on the field have applied the lithostratigraphic nomenclature suggested by Warrington *et al.*[9] The Permian (975m) at Wytch Farm comprises continental fluviatile-lacustrine deposits underlain by basal breccias. Continental sandstones and mudrocks also form most of the Triassic succession. The lower Sherwood Sandstone Group (c. 168m: this is within the Bunter Sandstone of the older nomenclature) consists of fine- to coarse-grained arkosic sandstones developed in fining-upwards units. Porosities range from 4% to 29%. Overlying the sandstones is the Mercia Mudstone Group (c. 360m: Keuper Mudstones), red brown mudrocks grading to grey claystones with thin limestones at the top. The thicker development of the Mercia Mudstone Group at Winterborne Kingston (Fig. 2.2) is partly due to the development of thick Saliferous Beds mentioned above.

The top of the Upper Marls on the coast section is represented by 15–20m of pale green silty marls (Tea Green Marls). These in turn are overlain by the first beds of the marine transgression which belong to the uppermost division of the Triassic, the *Rhaetic*. This is a fossiliferous shale interval in which thin bedded limestones are of increasing importance upwards in the sequence. The Rhaetic was encountered in the Portsdown well (Fig. 2.2) possibly in littoral facies, but was absent at Henfield and deep borings to the east. At Wytch Farm the uppermost 20m (Penarth Group) of the Triassic comprises micritic limestones overlying anhydritic mudstones.

A complete sequence of *Jurassic rocks*, about 1,200m thick, crops out along the Dorset coast. The succession follows conformably on the Rhaetic and can be divided into Lower (Lias), Middle (Dogger) and Upper (Malm) Jurassic. The *Lias* is in the order of 300m thick and consists of varyingly calcareous marine clays with limestones and intercalations of sandstones and shales. These sediments were deposited in a shallow sea within which Cornubia, western Brittany and most of Wales probably stood as islands, as did the Anglo-Belgian island stretching from the London Platform to the Ardennes. The Middle Lias in Dorset carries fine calcareous sandstones whilst the Upper Lias contains the important Bridport Sands (40m, thickening to 160m in the Winterborne Kingston borehole; Fig. 2.4). The Bridport Sands at the Wytch Farm Oilfield are about 70m thick and are probably slightly younger (Aalenian compared with mostly Toarcian) than at the coastal outcrops or in the Winterborne Kingston-1 borehole. The lithologies are bioturbated moderately-sorted coarse siltstones to very fine-grained sandstones probably deposited as a storm-dominated sheet sand in a transition zone to lower shoreface environments.[11] Porosities vary with the degree of calcite cementation between 10% and 32%, and the interval is one of the two producing horizons in the Wytch Farm Oilfield.

In France, the Lias crops out along the coast of the Contentin Peninsula. The sequence is dominated by fossiliferous limestones and calcareous shales, up to 160m or so thick.

The Lias was penetrated in all the deep boreholes along the south coast of England (Fig. 2.4) to the Brightling-1 hole. There is eastward thinning and overlap of the interval in Kent so that the Lias is represented by only a few

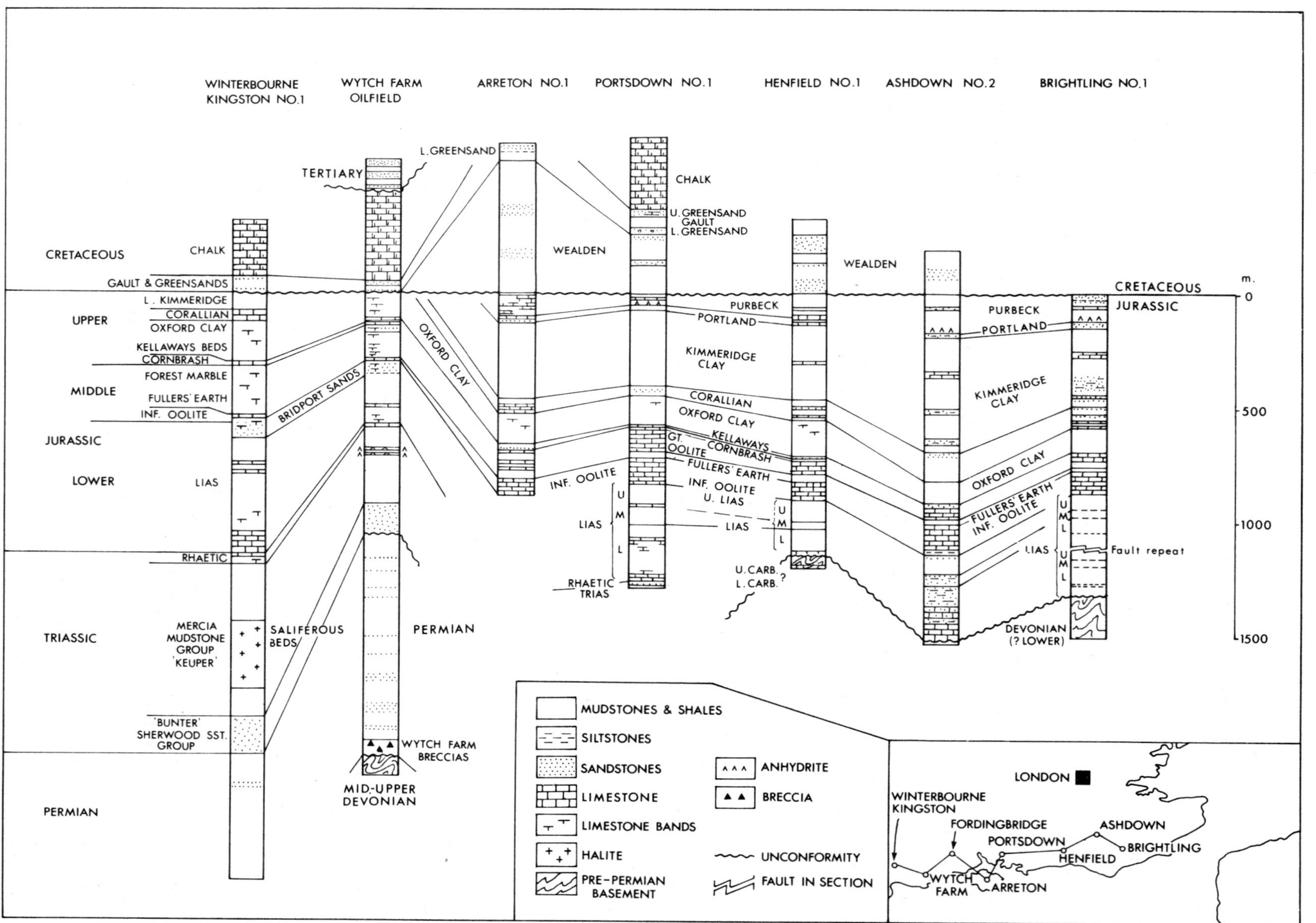

Figure 2.4. Correlations between selected deep exploratory boreholes in southern England (in the main from Falcon and Kent (1960),[10] and Colter and Havard (1981)[11]).

metres of rock at Harmansole and Dover on the coast, and is absent to the north on the London Platform.

The *Middle Jurassic* sequence in Dorset commences with the *Inferior Oolite* which is here a condensed (5m) section of limestones and marls. The same interval is extremely thick in the Cotswolds to the north and is represented in the Portsdown and Brightling wells by more than 100m of oolitic limestones.

The condensed nature of the Dorset sequence may reflect positive movement on the Weymouth Bay axis. Certainly the important Mendip Axis was active at this time, as demonstrated by the unconformable relationship of the Middle Jurassic and older rocks in that area, and the Cornubian massif may also have been further uplifted.

The Jurassic sea probably reached its maximum extent during deposition of the succeeding *Great Oolite Series* (Bathonian: Middle Jurassic). Despite its name, which derives from the thick development of the unit in north Somerset and the Cotswolds, the sequence is dominated by marine argillaceous shelf sediments with varying amounts of thin limestones. The interval is more than 100m thick in Dorset and is present in all the deep wells shown in Fig. 2.4. The thin marls and limestones of the Cornbrash are also widespread, marking another transgressive pulse, and they straddle the Middle-Upper Jurassic boundary.

The Bathonian in northern France (Bay of Seine) comprises a lower marly unit and an upper interval of bioclastic limestones. A composite Middle Jurassic thickness for northernmost France might be about 140m.

Thin shallow marine carbonates and clays are found at the base of the *Upper Jurassic* section (about 700m thick) in Dorset. These pass rapidly up into the *Oxford Clay*, which is a thick (150–180m) unit which extends across England and into the North Sea Basin. Bluish grey claystones with limestone nodules are the dominant lithology. The succession continues (Fig. 2.4) with the 60m thick limestones, claystones and sandstones of the *Corallian* formation, which represent shallower marine conditions. This is followed comformably by the important *Kimmeridge Clay* unit which ranges in thickness from 240m at the coast to almost 500m in the Kimmeridge area. It comprises dark grey marine claystones, mudstones and thin argillaceous limestones. Oil shales occur at intervals in the Kimmeridge Clay[13] and the unit as a whole is recognized as having high petroleum-generating potential. It is one of the main source rock intervals of the North Sea Basin.

The top of the Kimmeridge Clay is marked by shallowing marine conditions. Along the south coast of England the upper part of the Jurassic is represented by a complex of shallow marine to lagoonal clastics, carbonates and evaporites, the *Portland Beds* and *Purbeck Beds*. The boundary between the Jurassic and the Cretaceous falls within the Purbeck Beds. The two units comprise more than 200m of varied lithologies representing gradually shallowing conditions. In the Purbeck Beds lacustrine and lagoonal deposits became important, with fossil soils and freshwater limestones in the upper part.

The Upper Jurassic sequence of the coastal exposures continues in the subsurface of southern England (Fig. 2.2). The section in the Portsdown well is somewhat thinner indicating the presence of a positive element – the Portsdown Swell – separating the Wealden and Hampshire sub-basins at this time. Westwards the section above the Corallian is cut out by erosion on the north side of the Purbeck-Isle of Wight line, for example at Wareham and Wytch Farm.

Upper Jurassic beds crop out in the western part of the Bay of Seine. The sequence there is not unlike that in southern England, with many of the main units being represented. The Kimmeridgian, for example, is again dominantly argillaceous and occasionally bituminous or lignitic.

There are extensive outcrops of *Cretaceous* rocks on both sides of the Channel; from Dorset to Kent in England and from Le Havre to Calais in France. An obvious lithological division is present between a lower sand and shale interval and an upper limestone (Chalk) interval. The base of the Cretaceous is actually within the Purbeck, above which the sequence continues as follows:

3. Gault Clay and Upper Greensand
2. Lower Greensand
1. Wealden.

The *Wealden* reflects continued marine regression. Rhythmic lacustrine and deltaic shales, siltstones and sandstones dominate a complex sequence. In the Weald of Kent the interval may be 700m thick, but thins away from this area. However, there is thickening also in the Isle of Wight – Swanage area (600m in the Arreton-2 well) which may represent depositional variation across the Purbeck-Wight line accentuated by later erosion. As might be anticipated in a non-marine deltaic lacustrine complex, there is considerable lateral variation in facies.[14] An overall westward coarsening suggests that the Cornubian landmass was a prime source for detritus. An increasing marine influence is evident at the top of the Wealden sequence.

The *Lower Greensand* was deposited following an important period of tectonic uplift and erosion (Late Cimmerian). The transgressive marine sediments of the Lower Greensand rest unconformably on beds as old as the Oxford Clay (see Wytch Farm on Fig. 2.4) but in the east the passage is more generally conformable. Shallow marine sandstones with claystones dominate the sequence. About 200m of Lower Greensand is present in the western Weald and a similar thickness is present on the Isle of Wight, but the interval thins westwards and northwards, being thin at Portsdown-1 (27m) and absent at Lulworth and in the Bere Regis well (Fig. 2.3). There was renewed transgression late in Early Cretaceous time so that, to the west, beds of this age progressively overstep older strata and rest on Triassic rocks in Devon.

On the Isle of Wight the succession continues with about 30m of blue grey clay (*Gault Clay*) which passes upwards by gradation into a similar thickness of glauconitic sandstones (*Upper Greensand*). The Gault Clay becomes thinner and sandier westwards, and eventually becomes indistinguishable from the overlying sandstones. Both units thicken eastwards from the Isle of Wight into the western part of the Weald but thin again in east Sussex and Kent. Equivalents of part of the interval on the French coast are represented by green glauconitic clays.

The *Upper Cretaceous Chalk* represents a widespread regional transgression – the Cenomanian transgression – which affected wide areas of the North Atlantic borderlands, including the North Sea Basin. A widespread blanket of fine-grained limestone—chalk—was deposited in the clear detritus-free seas. The rock is composed of microscopic algal material (coccoliths) with varying amounts of other shell material. The Chalk of the Hampshire Basin is more than 500m thick whereas on the French coast only 150–200m are present. At the end of chalk deposition there was widespread regression of the sea and a period of gentle folding and erosion (Laramide fold phase) before deposition of Tertiary strata.

The *Tertiary* in southern England is represented by an incomplete and varied sequence of sediments. *Lower Tertiary (Palaeogene)* strata up to 650m thick are preserved in the Hampshire Basin – Isle of Wight area. They were deposited in a marine arm of the North Sea Basin and comprise a cyclic sequence of sandstones, claystones and mudrocks representing fully marine to coastal plain environments. During the Oligocene the basin became smaller and more restricted, with fewer marine incursions. Sands, clays, marls and thin lignites were deposited in lagoonal and freshwater conditions. A similar upward passage from marine to non-marine rocks is also present in the Tertiary of the Paris Basin. Younger Tertiary (*Neogene*) sediments have only scattered minor development on the margins of the Channel Basin.

An isolated outlier of Tertiary rocks is found in two fault basins, near Bovey Tracey and at Petrockstow in Devon. More than 200m of clays and gravels with interbedded lignites have been penetrated in borings and probably represent erosional debris from the nearby granite batholith which collected in freshwater lakes during Oligocene time.

Offshore Stratigraphy

We turn now to a consideration of the stratigraphy of the offshore areas of the Channel Basin, as deduced from sampling and shallow seismic work. Comments on the deeper subsurface portions of the basin, must inevitably be speculative.

Basement (Pre-Permian) rocks are restricted in the main to nearshore extensions of onshore outcrops. There is a large area of Basement rocks, comprising phyllites, schists and granites, off Cherbourg (Fig. 2.1). Devonian and Carboniferous rocks also form a wide outcrop band around Land's End.

Permo-Triassic rocks (at least 300m thick) have been mapped in a broad outcrop from the Lizard to Start Point and eastwards around Lyme Bay as a natural extension of the onshore occurrences of the same rocks. Red and green sandstones, mudrocks and marls have been retrieved but are unfossiliferous so that the Permo-Triassic age assigned is based entirely on lithology.

Triassic red and variegated conglomerates, sandstones and claystones are also found offshore from the west coast of the Contentin Peninsula, where they are less than 40m thick and rest with marked unconformity on old Basement rocks.

The depositional basin of erosional redbed

sediments extended from the south coast of England northward into Somerset and southeastwards under the English Channel. The eastward limit of the basin is unknown. The alluvial fans are regarded as a marginal facies to a desert basin which lay to the east. Within this desert basin, sands and silts of fluviatile origin and wind-blown dune sands were deposited contemporaneously.

The basal breccias pass eastwards below the Channel Basin into sandstones and siltstones and are probably similar in many ways to the Rotliegendes deposits of the Southern North Sea Basin. Because of the nature of their deposition, the sandstones are likely to include porous, potential hydrocarbon reservoir rocks, the prospectivity of which will be dependent upon their relationship with either older or younger hydrocarbon source rocks.

Jurassic rocks crop out on the seabed in Lyme Bay and the Bay of the Seine as extensions of the onshore outcrop bands. There are also separate broad outcrop belts in mid-Channel between Cherbourg and the Isle of Wight (Fig. 2.1). Sampling has shown the offshore *Lower Jurassic* section to be dominated, as are the onshore sections, by shallow marine carbonates. Biosparites and shelly biomicrites alternate with more argillaceous horizons. The *Middle Jurassic* limestone samples offshore are again comparable to their coeval onshore counterparts, although less exactly so in the lower part of the section. It has been pointed out that the presence of Liassic sediments south of Start Point and Middle Jurassic rocks north of the Channel Islands indicates that the Start – Cherbourg Ridge was submerged at this time, although the stratigraphic units may thin onto the axis. Most of the *Upper Jurassic* lithostratigraphic units of the onshore sections have been sampled offshore. The Kimmeridgian for example consists of dark grey bituminous shale with shelly, biomicrites and marls comparable to the interval onshore. It is probable that the Jurassic underlying the Channel Basin is similar in thickness to the known onshore sections, that is in the order of 1,000m.

The twofold subdivision of the *Cretaceous* persists throughout the Channel Basin. *Lower Cretaceous* rocks rest unconformably on the Jurassic and crop out widely offshore; in a broad area off the Isle of Wight, in an east-west outcrop belt in mid-Channel north of Contentin, and in a continuous band between southeast England and the Boulonnais coast of France. Probable equivalents of both Wealden and Greensand facies have been sampled. Sandstones, conglomerates, mottled claystones and marls are found in a variable Wealden sequence (Boulogne region and south of Isle of Wight) which may be up to 400–500m thick. The overlying Greensands consist of glauconitic argillaceous siltstones and fine-grained sandstones with coarser sand and carbonaceous intercalations. The Gault Clay is absent over the northern part of the basin with the result that the Upper and Lower Greensands cannot be differentiated. Where present the Gault is black, clayey, micaceous and glauconitic, and contains a rich fauna of foraminiferids.

The *Chalk* (Upper Cretaceous) rests unconformably on the Lower Cretaceous and older beds and is disposed in broad outcrop bands which pick out the major offshore structures (Fig. 2.1). Basal glauconitic and marly intervals are followed upwards by a normal range of chalk lithologies with flints. The Upper Cretaceous of the central and eastern English Channel is between 250–300m thick.

The main *Tertiary* outcrop in the Channel Basin is an extension of the Hampshire Tertiary Basin onshore in England, and lies within the eastern or Dieppe sub-basin of the Channel.

The Tertiary sequence exposed in the Hampshire Basin and on the Isle of Wight includes sediments of both Eocene and Oligocene age and reaches a cumulative thickness of 600m. This Lower Tertiary basin extends southeastwards across the Channel where it forms a broad downfolded tract between the coasts of Sussex and Kent and that of northern France (with a thickness of about 650m). No sediments younger than Upper Eocene age have been proved by the bottom sampling programme over this eastern part of the Channel Basin.

The Tertiary sediments consist of a broadly cyclic sequence of sandstones and shales with subordinate limestones, probably reflecting depositional environments ranging from fully marine to lagoonal, freshwater and continental. Onshore in southeastern Britain, the Oligocene sediments have been involved in a phase of strong mid-Tertiary folding, related to the major Alpine movements of southern Europe. There is also a well-defined mid-Tertiary unconformity between rocks of Miocene and Eocene age over the western part of the Channel Basin, even though the Tertiary sequence there is more fully marine.

It is not known whether the western and eastern section of the Channel Basin were physically separated during the Tertiary or whether they formed a single depositional basin which had a stronger marine influence towards the west. It seems probable that the two sub-basins became separated during the mid-Tertiary Alpine movements, because separation of the Hampshire and Paris Basins also occurred at that time.

Hydrocarbon Potential

The Lower Palaeozoic and older rocks beneath the Channel Basin have low potential as source or reservoir rocks. The reservoir properties of Upper Palaeozoic rocks were similarly diminished by the Hercynian (end Carboniferous) orogeny. Westphalian Coal Measures, known in the subsurface in Kent, are probably potential gas source rocks, although their offshore extent may be limited.

Thick Permian and Triassic sandstones, which are productive in the Southern North Sea Basin and at Wytch Farm, also occur beneath the Channel Basin. Porosities range from fair to good and there is little doubt that juxtaposed against or above adequate source rocks the thick sandstones could provide productive reservoirs.

The thick Jurassic section of the Channel Basin certainly contains interesting source and reservoir horizons. Potential reservoir horizons exist in sandstones of Middle and Upper Liassic age, and in limestones of the Middle Jurassic, while thick marine clays and shales of Liassic and Upper Jurassic (Oxford Clay and Kimmeridge Clay) age provide excellent potential source rocks. Active oil seeps are known in several localities along the Dorset coast and traces of both oil and gas have been recorded in a number of wells. Butuminous veins are common in the Lias whilst the Kimmeridge Clay contains a number of oil shale horizons.[13]

The extent of Liassic sandstones in the offshore is not known in detail. Similarly the bioclastic and oolitic Liassic limestones of the French coastal sections may also provide prospective reservoirs beneath the Channel Basin but their precise development is unknown.

Although a number of wells have been drilled in southern England to test the Jurassic prospects, only three oilfields have been discovered to date; Kimmeridge, Wareham and Wytch Farm. The Kimmeridge Oilfield (discovered in 1959) on the Dorset coast (Figs 2.3 and 2.5) produces from a fractured Middle Jurassic marine limestone (Cornbrash) with minor production from the overlying fractured shale (Oxford Clay). The Kimmeridge Oilfield is developed in a minor fold on the steep north limb of the larger Weymouth Bay anticlinal structure. Wareham (1964) produced oil and water (declining from 100 barrels per day combined) from the 5m thick Inferior Oolite (Middle Jurassic) and the top 3m of the Bridport Sands. The field began production in 1970 but was shut-in in 1979. Drilling in the early 1970s however, produced a significant discovery at Wytch Farm, a few miles northeast of Kimmeridge. The Wytch Farm Oilfield[11] was discovered in 1973 by the Gas Council/BP grouping. The original producing horizon was in the Liassic Bridport sands. However, a deeper well on the same structure in 1977 found substantially more oil in Triassic sandstones (Sherwood Sandstone Group). Structurally the field is an east-west trending tilted fault block controlled by a fault to the south, with dip and minor faults providing closure on the other flanks (Fig. 2.5). Wytch Farm lies a short distance north of the important Purbeck – Isle of Wight Fault Zone which had a strong influence on stratigraphic development in the area (see for example the difference between the Wytch Farm and Arreton wells in Fig. 2.4). The oil is thought to have its source in Jurassic sequences south of the Purbeck – Isle of Wight line, which are known from analysis of samples of the Arreton-2 well to have high potential and to be mature, in contrast to rocks north of the fault. A phase of pre-Upper Cretaceous faulting gives rise to a strong down-to-the-south movement on the Purbeck Fault resulting in deeper burial of marine source rocks in the south and erosion of some potential source beds (Kimmeridge Clay) to the north. Tertiary (Alpine) folding later gave rise to the northward facing monocline along the Purbeck line. Generation and migration of the oil is assumed to have occurred between the two tectonic episodes. Oil production from Wytch Farm started in March 1979 at a rate of 1,000 b/d, with plans to increase this towards 15,000 b/d.

Proved reserve estimates at Wytch Farm are in excess of 100 million barrels and recently figures of double this quantity have been quoted. In addition to the Wytch Farm structure there are a number of other prospective structures in the area and an interesting oil discovery has been made by the Candecca group on the Humbly Grove structure. It is probable that more oil remains to be found in the same general area.

In addition to the producing fields a number of significant hydrocarbon shows have been en-

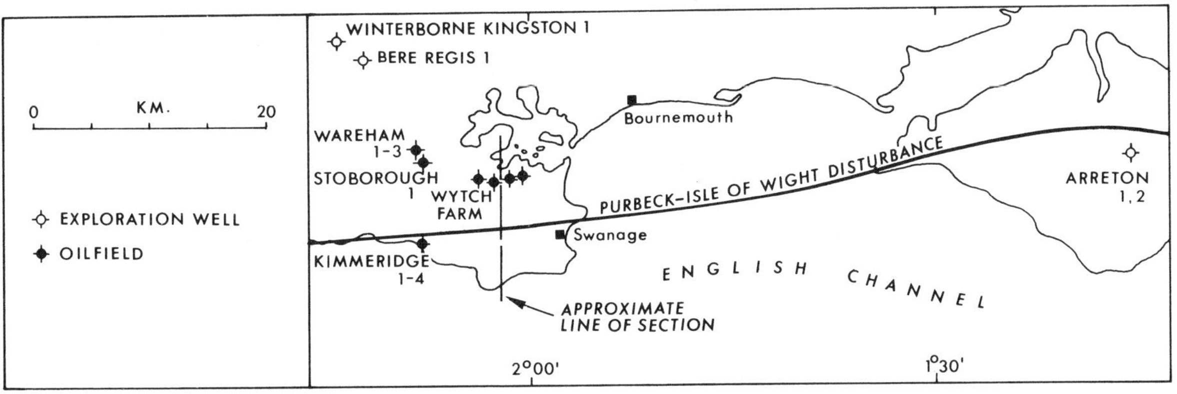

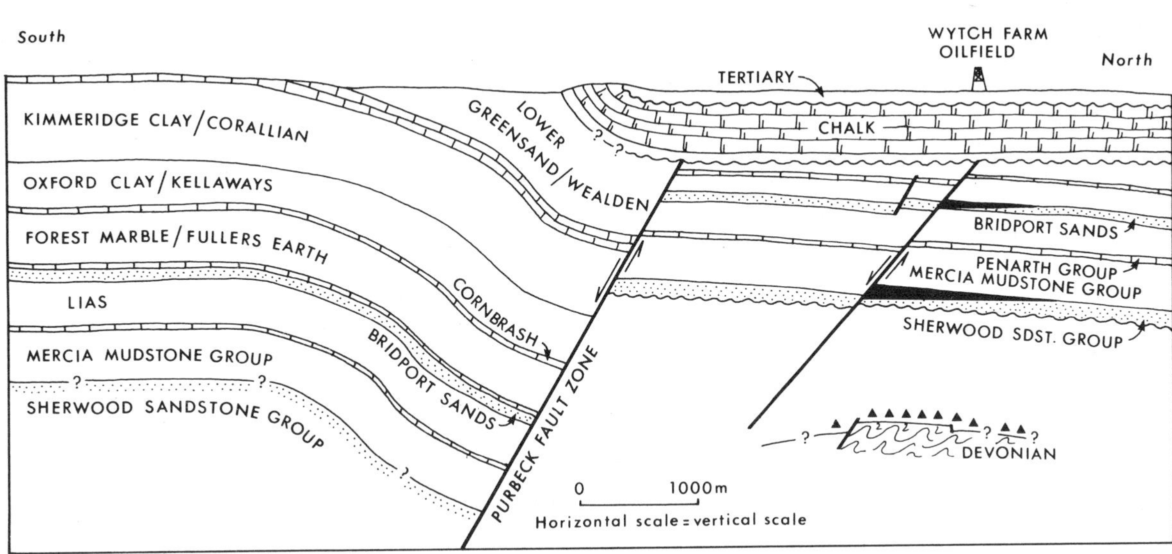

Figure 2.5. Cross-section and location map of Wytch Farm Oilfield and Purbeck Fault Zone (with modifications after Colter and Havard (1981)[11]).

countered in other onshore wells. The basal sandstones of the Upper Jurassic had oil shows in the Brightling-1 and Ashdown-1 wells, and the latter also had gas shows in Upper Jurassic Corallian Limestones and Purbeck Beds. The Purbeck also yielded subcommercial gas in BP's Heathfield hole, although an 1896 water well drilled nearby flowed methane which was used to light the railway station for half a century. The Portland Beds produced traces of gas in Brightling-1. In addition Esso made an interesting gas discovery at Bletchingly in Surrey before pulling out of onshore exploration in the late 1960s, and this discovery remains to be properly assessed.

The thick sandstone intervals of the Lower Greensand and Wealden offer obvious potential reservoirs. Intercalated marine argillaceous units provide good source material. The interval is productive in the Kinsale Head Gas Field of the Celtic Sea (Chapter 4), and has oil seeps and possible gas shows in the area of the Weald. There is oil production from the Lower Cretaceous in the West Netherlands and also in the Lower Saxony Trough. Variable porosity and complex facies patterns, particularly in the Wealden, are factors which mitigate against the Lower Cretaceous as an exploration target. Furthermore the Lower Cretaceous is at shallow depths over much of the Channel Basin and is exposed in many of the structures. The potential of the horizon increases as the Chalk and Tertiary cover thickens westwards.

The Chalk is widely exposed or at shallow depth throughout the basin; this combined with poor permeability gives the interval a poor exploration potential for source and reservoir rocks. The shallow depth of burial and generally synclinal disposition also render them non-prospective within the Channel Basin.

Systematic onshore oil exploration in southernmost England dates back to the work of the D'Arcy Exploration Company (precursor of British Petroleum) in the 1930s. Until the end of the 1960s BP was the dominant company operating in the region, although other companies (notably Esso) were also active. Over this period a number of deep exploration boreholes were drilled in the Weald and Hampshire Basins[10] which considerably advanced knowledge of the subsurface stratigraphy (some of these wells are shown on Fig. 2.3). Minor oil and gas shows were encountered in a number of these as mentioned above. The Kimmeridge Oilfield was discovered by BP in 1959 and Wareham in 1964. Production however was small and the onshore exploration results at this stage had been disappointing due to the lack of adequate reservoir rocks. Nevertheless new companies entered the search during the 1970s but it fell to the Gas Council – BP Group to discover the Wytch Farm Oilfield in 1973. This discovery has helped to stimulate the onshore exploration search but has also provided an optimistic pointer to the offshore Channel Basin.

It is perhaps surprising in view of later events that the first offshore exploration well in British waters was in the English Channel. The well, BP Lulworth Banks No. 1 (see Fig. 2.3) was drilled in 1963, 13 km west of the Kimmeridge Oilfield to a depth of 762m, but was abandoned as a dry hole. However, the Channel area is one of the world's busiest seaways and this makes oil exploration activities hazardous. In addition the median line between France and Britain was in dispute and this continued for some years until settled by Arbitration in 1977. The caution and uncertainty caused by these two factors has delayed offshore exploration for petroleum in the Channel Basin, compared with other offshore basins. The first speculative regional seismic survey designed for the oil industry was shot in the Channel in 1972 by Seiscom Delta. There has been steady acquisiton of data by the industry since that time. The UK Fifth Round of licensing in 1977 included for the first time the issue of two blocks in the mid-Channel area, one to Hydrocarbons GB Ltd and the other to Conoco. In 1978 Hydrocarbons GB spudded the 98/22-1 well, later junked, but 98/22-2 was later abandoned as a dry hole. The Department of Energy issued further exploration licences covering parts of the Channel to a range of companies in the Seventh Round of Licensing. One of these allocations was of Block 98/11 to British Gas Corp. – BP which lies on trend from the onshore Wytch Farm Field. The coming years will see the testing of the Channel Basin's petroleum potential by further deep offshore exploration wells.

Early in 1982 the French Government finally licensed six large (varying from approximately 600 sq km to 1200 sq km) areas along the Channel eastwards from Cherbourg to Dieppe to a number of French and foreign oil companies. Increased exploration activity can be expected in the area during the next few years.

References

1. KING, W. B. R., 1949. The geology of the eastern part of the English Channel. *Q. J. geol. Soc. London* **104**, 327–337.
2. DONOVAN, E. T. and STRIDE, A. H. 1961. An acoustic survey of the sea floor of Dorset and its geological interpretation. *Philos. Trans. R. Soc. London* **B. 244**, 299–333.
3. DINGWALL, R. G. 1971. The structural and stratigraphical geology of a portion of the eastern English Channel. *Inst. geol. Sci. Rep.* 71/8, 24 pp.
4. DINGWALL, R. G. and LOTT, G. K. 1979. IGS boreholes drilled from M. R. Whitethorn in the English Channel 1973–1975. *Inst. geol. Sci. Rep.* 79/8, 45 pp.
5. BUREAU DE RECHERCHES GÉOLOGIQUES ET MINIÈRES 1974. Geological Map of the Continental Margin: English Channel. Scale 1:1,000,000 with *Explanatory Notes.*
6. SMITH, A. J. and CURRY, D. 1975. The structure and geological evolution of the English Channel. *Philos. Trans. R. Soc. London* **A279**, 3–20.
7. LARSONNEUR, C., HORN, R. and AUFFRET, J. P. 1975. Geologie de la partie meridionale de la Manche centrale. *Philos. Trans. R. Soc. London* **A279**, 145–154.
8. HAMILTON, D. 1979. The geology of the English Channel, South Celtic Sea and Continental Margin, South Western Approach: *in* Banner F. T., Collins, M. B and Massie, K. S. *(Eds). The North–West European Shelf Seas: 1. Geology and Sedimentology.* Elsevier, 61–87.
9. WARRINGTON, G., AUDLEY-CHARLES, M. G., ELLIOT, R. E., EVANS, W. B., IVIMEY-COOK, H. C., KENT, P. E., ROBINSON, P. L., SHOTTON, F. W. and TAYLOR, F. M. 1980. A correlation of Triassic rocks in the British Isles.

Geological Society of London, Special Report No. 13, 78 pp.

10. FALCON, L. N. and KENT, P. E. 1960. Geological results of petroleum exploration in Britain 1945–1957. *Geological Soc. London, Memoir No. 2*, 56 pp.
11. COLTER, V. S. and HAVARD, D. J. 1981. The Wytch Farm Oil Field, Dorset: *in* Illing L. V. and Hobson G. D. (*Eds*). *Petroleum Geology of the Continental Shelf of North-West Europe.* Heyden & Son Ltd. London, 494–503.
12. DINGWALL, R. G. 1977. (Compiler). Winterborne Kingston No. 1 Geological Completion Report. *Inst. of Geological Sciences, Limited distribution report.*
13. GALLOIS, R. W. 1978. A pilot study of oil shale occurrences in the Kimmeridge Clay. *Inst. of Geol. Sciences Report* 78/13, 26 pp.
14. ALLEN, R. 1981. Pursuit of Wealden models. *J. geol. Soc. London* **138**, 375–405.

Chapter 3

Western Approaches Basin

Introduction

The Western Approaches Basin is a deep structural trough which follows the line of the English Channel and the Western Approaches between the basement massifs of Southwest England (the Cornubian Massif and its offshore subsurface extension to the west-southwest) to the north, and Brittany (the Armorican Massif) to the south. To the east the basin is separated from the Channel Basin by a cross-cutting barrier of basement rocks extending from Cherbourg to the southern tip of Devon—the *Start-Cherbourg Ridge* (Fig. 3.2). The western margin, on the outer part of the shelf, is defined by a positive northwest-trending basement element—a shelf edge feature seen in several of the basins described in this volume. The east-northeast to west-southwest trend of the Western Approaches, the Channel and Celtic Sea Basins contrasts with the northeast or north-northwest trend prevalent in other basins on the continental shelf west of Britain and Ireland.

The Western Approaches trough is thought to have been initiated as a Late Palaeozoic-Early Mesozoic (Permo-Triassic) rift structure subsequently infilled with thick Permo-Triassic, Jurassic, Cretaceous and Tertiary sedimentary sequences. The southern margin is strongly faulted against the Precambrian and Palaeozoic Armorican Massif, while the northern margin also shows some well developed local faulting.

In general terms water depths in the Western Approaches increase from east to west. The one hundred metre isobath trends from the Isles of Scilly southeastwards to the Isles d'Ouessant off the Brest peninsula. The more restricted area east of this isobath is covered by water depths ranging between 60m and 100m. A striking feature is the linear closed 100m contour trending west-southwest which outlines the western extension of the Hurd Deep. West of the Scilly-d'Ouessant isobath there is a broad shelf which deepens gradually to 200m at the shelf edge before plunging off down the Continental Slope to depths of 4,000m. The Slope here is markedly incised by a number of deep steep-sided, slightly sinuous submarine canyons, in contrast to the Goban Spur area immediately to the north.

Geophysical Data

Early geological work in the Western Approaches and western English Channel area was carried out by King and summarized by him in a 1954 publication.[1] In 1957 a team from Bristol University began a systematic study of the area using a basic bottom sampling grid of 10 minutes longitude and 10 minutes latitude, but employing a much closer spacing in areas of geological complexity. About 60% of the 1,600 stations yielded useful information, and some 2,400 nautical track miles of continuous seismic profiling were collected.[2,3]

The geological work together with early seismic reflection[4] and refraction data estab-

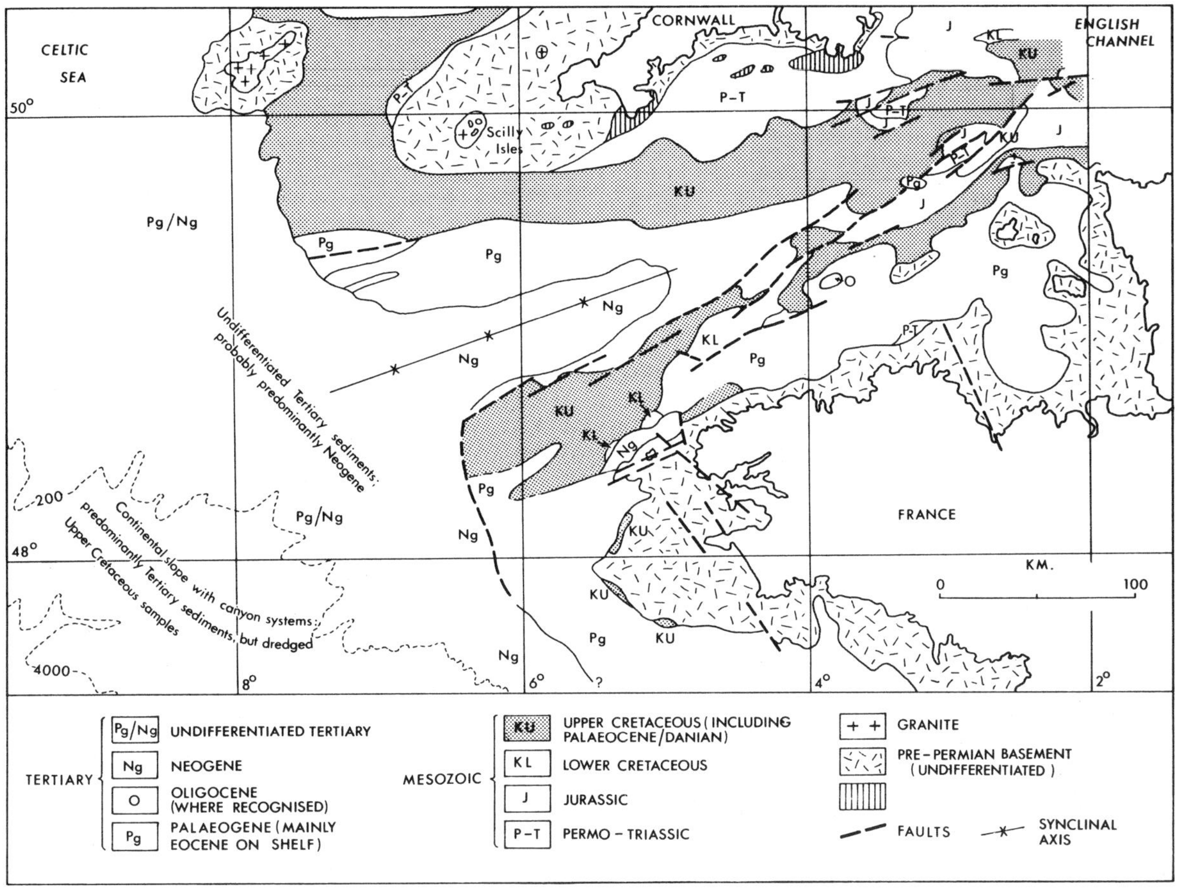

Figure 3.1. Sketch map of the solid geology of the Western Approaches (in the main after Hamilton (1979)[12] and B.R. G.M. (1974)[11]).

lished the existence of an elongate trough in the region which was assumed to be filled with about 3.5km of Permo-Triassic to Recent sediments.

Regional gravity[5] and magnetic coverage from a variety of sources is also available. The magnetic anomalies[6,7] for the most part trend parallel to the basin axis but on the outer part of the Shelf they are parallel to the Continental Slope. The faulted southern margin of the basin out to the edge of the Shelf is very well defined on the magnetic maps.

Published seismic profiles demonstrate the existence of two discordant structural units in the Western Approaches Basin. The upper unit is formed of essentially flat-lying or gently folded sediments which rest unconformably on the eroded surface of a strongly folded and faulted lower unit. An analysis of seismic velocities over the basin suggest that the upper unit is composed of a Permo-Triassic to Tertiary sequence, while the lower unit shows markedly higher velocity values and is believed to represent folded and metamorphosed Palaeozoic basement rocks. Deep hollows in the basement surface occasionally contain a thin high velocity layer, which could be either basal Permian or uppermost Carboniferous strata.

Three seismic velocity layers have been recognized:[8]

3. *Upper sequence* (1.9–3.6 km/sec) with acoustic unconformity at base; of probable Cretaceous-Tertiary age.
2. *Intermediate sequence* (3.6–4.3 km/sec) Permo-Triassic (and Jurassic?) with possible Lower Cretaceous at the basin margins.
1. *High velocity, deep sequence* (4.4–6.6 km/sec) which represents metamorphic basement and Palaeozoic rocks.

The maps shown on Fig. 3.2 summarize the main results drawn from the seismic exploration. They show that the Western Approaches Basin comprises a number of subsidiary depositional centres separated by northwest-trending positive ridges. The influence of the eastern bounding *Start-Cherbourg Ridge* is clearly seen. Both this ridge and the similar feature which runs from the Lizard to Marlaix and separates the two eastern sub-basins, are probably complex and may contain intrusive centres as well as faulted basement rocks. These two ridges, particularly the Start-Cherbourg Ridge, exerted a considerable influence on sedimentation and marine transgression during Mesozoic time. Further west a broad northwest-trending positive ridge forms the outer part of the continental shelf, while on its seaward side a westward-thickening prism of sediments comprises the *Continental Edge Basin*.

The southern margin of the Western Approaches Basin consists of the seaward extension of the *Armorican Massif* as the *South Armorican Continental Shelf*, which has a relatively thin cover of younger sedimentary rocks. In the north the margin comprises the Cornubian-Scilly extension, formed by a series of Hercynian granite batholiths, much as in Cornwall, which plunge gently towards the continental margin and have been detected by sampling and geophysical methods.

The seismic reflection surveys have shown considerable faulting within the basin; some faults affect only basement rocks whilst others cut younger sediments to varying degrees. Fault-controlled basement ridges run parallel to the basin axis and are particularly well-defined west of 6°W. There are also offsetting sets of northwest-southeast and north-northwest-south-southeast faults.

Stratigraphy

The *Pre-Permian basement rocks* which underlie the complex downfaulted trough of the Western Approaches Basin are believed to be similar to those which flank the trough, and which form the Cornubian Peninsula to the north and the Armorican Massif to the south. Devonian and Carboniferous rocks have been found by offshore sampling, rimming the Cornubian Peninsula. Onshore the thick Devono-Carboniferous 'Culm' sequence of sandstones and shales are metamorphosed and folded. Offshore they comprise cleaved dark shales, slates and some phyllites, which have not yielded fossils. However, carbonized spores have been found in Devono-Carboniferous rocks near Haig Fras. The Haig Fras Granite itself crops out on the seabed and has been sampled and mapped in detail by sparker surveys. Samples have yielded an age of 275 ± 10 My (Late Carboniferous).

Seabed sampling[2] has also revealed submerged extensions of the rocks of the Channel Islands and of the Lizard-Start Complex. The rocks found off Start Point bear close resemblance to the onshore ultrabasic-schist-granite complex, with pale schists predominant. Granitoid gneiss is found at four localities off Plymouth and emerges at one point to form the Eddystone Rocks (375 ± 17 My; Devonian). The basement rocks of Brittany, which extend offshore, comprise a range from granites to gneisses together with metamorphosed and unmetamorphosed Palaeozoic sedimentary rocks.

In Brittany, the Palaeozoic consists of Ordovician quartzites, Cambrian phyllitic sandstones and arkoses, and sandstones, schists, shales and mudstones of uncertain age (Cambrian to Early Devonian). Precambrian rocks crop out mainly in the Contentin Peninsula and consist of sericitic phyllites and schists locally associated with black cherts. Massive granitic batholiths (e.g. Barfleur Granites) are intruded into the Precambrian. The Palaeozoic rocks of Brittany are highly tectonized, granitized and interfolded with true Precambrian basement.

The following account of the post-Carboniferous stratigraphic sequence is based on the work of many geologists, some of whom have written papers which summarized knowledge at a particular time.[1,2,9–12] As elsewhere in the book, the reference list accompanying this chapter is not intended to be comprehensive, but to serve as a guide to the more important papers in the literature.

During *Permo-Triassic* times the fault-controlled basins southwest of the British Isles became centres of deposition of thick redbed and evaporitic sequences derived from the progressive erosion and destruction of the young Hercynian mountain chain in a desert environment. Deep valleys draining from the surrounding uplands converged into the Western Approaches and Channel Basins, while across the floors of the basins themselves semi-arid and desert conditions prevailed with sands and pebble beds being deposited both as dunes, and in outwash delta fans and flood channels. This pattern of continental deposition continued until Upper Triassic time, when the original Hercynian highlands had become so worn down that only finer grained sediments such as

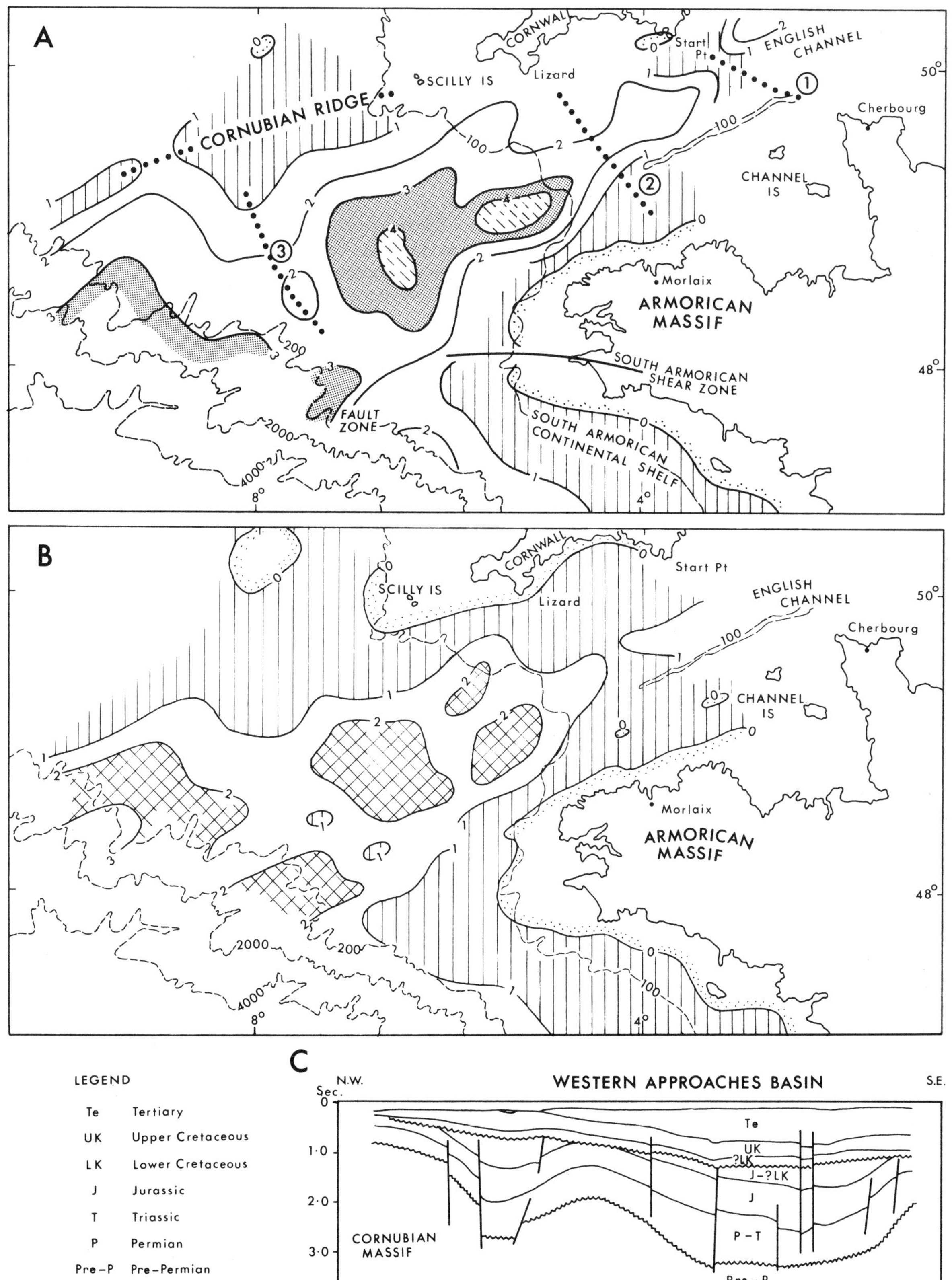

Figure 3.2A. Schematic structure contours of the metamorphic basement (depth in km) 1. Start – Cherbourg Ridge 2. Lizard – Marlaix axis 3. Outer Shelf axis. (Simplified after Avedik (1975)[8]).

Figure 3.2B. Isopachytes (in km) of post-Palaeozoic sediments including Permo-Triassic (simplified after Avedik (1975)[8]).

Figure 3.2C. Interpretive section based on a seismic profile across the Western Approaches Basin (after Naylor and Mounteney (1975)[15]).

river-deposited red mudstones and siltstones, and lacustrine salts and evaporites were being laid down across the lowland plains. Salt deposits are known to exist widely in the Triassic beds throughout Britain, and, within the vicinity of the Western Approaches, have been found in Somerset, in the Cheshire Basin, and as indications on geophysical records across the Celtic Sea and Southern Irish Sea.

The Western Approaches Basin is fringed by exposures of Permian and Triassic sediments. To the north, around the margins of the Cornubian Massif, these form a redbed sequence which passes conformably upwards into the Jurassic sediments. Further south a similar sequence containing a number of salt horizons occurs in Portugal. To the west, salt of Late Triassic to Middle Jurassic age has also been recorded in wells on the east Canadian Continental Shelf. It therefore seems likely that a thick Permo-Triassic redbed sequence, possibly containing salt horizons, will occur at depth in the Western Approaches Basin.

The Permo-Triassic sequence exposed along the coast of South Devon is about 2,750m thick and rests unconformably on folded older rocks. There is no palaeontological evidence to aid in the location of the Permian-Triassic boundary or, indeed to confirm the existence of undisputable Permian rocks.[13] The succession fines upwards from breccias and conglomerates near the base to red sandstone-siltstone-marl alternations in the upper part. Permo-Triassic red beds also crop out in mid-Channel northwest of Guernsey and on the north side of the Hurd Deep. Possible Permo-Triassic rocks have also been identified to the north of Brittany. The Permo-Triassic rocks of the Western Approaches region are entirely of continental redbed type. It seems that the basement ridges at the east of the basin prevented marine invasions into the continental New Red Sandstone basin.

At the end of Triassic times (Rhaetic) the sea is thought to have invaded the downfaulted depression from both the north and southwest. Renewed marine invasion in the Lower Jurassic may have come from the south by way of the Paris Basin. It is probable that the entire Western Approaches area, including the adjacent land areas of France and southern England became submerged beneath a partially enclosed shallow sea in which Jurassic sandstones, shales and limestones were deposited. In the Western Approaches it is likely that thick limestones associated with sandstones occur towards the shelf margin (west of 8°W), as they do in the Aquitaine Basin of France. Throughout long periods of the Jurassic, the *Brabant Massif* formed a positive island feature which extended from London into northern France, although it may have been submerged during Liassic and Kimmeridgian times.

Jurassic rocks have a restricted outcrop within the Western Approaches Basin (Fig. 3.1) but are interpreted from seismic records as having a wide distribution in the subsurface (Fig. 3.2C). Postulated subcrop maps at the sub-Chalk[12] unconformity in the eastern part of the basin show an ascending stratigraphic sequence from Triassic to Lower Cretaceous between the Lizard and Morlaix in broad subcropping bands.

The known outcrops in the east of the basin reveal that the Lower Jurassic comprises dark grey clays and pale calcilutites whilst the Middle Jurassic is represented by pale grey marine clays, calcarenites and sandstones. Upper Jurassic rocks have not been recorded at seafloor outcrop. Liassic shales and calcilutites with gypsum occur in a small inlier at 4° 20′W and this is the most westerly seafloor outcrop of Jurassic strata. Estimates in the order of 1,000m have been made for the thickness of the Jurassic in the eastern part of the basin whilst geophysical evidence suggests that more than 1,500m may be present further west.

The Jurassic stratigraphy of the sections in southern England has been discussed in the previous chapter. It can be anticipated that the stratigraphy of the Western Approaches will bear some similarity to the onshore sections but the western part of the basin is well removed from the onshore control and it is necessary to turn to the Celtic Sea and Eastern Canadian offshore for guidance (Chapters 4 and 10).

Figure 4.5 shows the sequence established within the Bristol Channel, and when compared with the Jurassic sequence of the south coast of England, shown in Figure 2.4, it can be seen to differ markedly in its conspicuous lack of any carbonate section. Although the Jurassic sequence in the originally adjacent areas of Portugal and the Grand Banks are characterized by thick carbonate sections, it is uncertain whether similar thick carbonates typify the Jurassic in the Western Approaches Basin. The available seismic reflection data suggest that in the central part of the basin a thick argillaceous sequence comparable to that found to the north in the Bristol Channel

Basin, and in the Mochras Borehole on the edge of the Cardigan Bay Basin, may be more likely. However, if carbonates are present, possibly towards the margins of the basin, they could act as good potential reservoir horizons for the accumulation of hydrocarbons. Onshore in southern England, good source rocks, in the form of black, organic-rich shales are present within the Lower Jurassic section, and similar horizons may extend offshore.

Unconformities and non-sequences are known in the onshore sequences in England, and similar breaks might be anticipated offshore. The positive Mendip Axis onshore was active at the beginning of Middle Jurassic time so that Middle Jurassic strata are unconformable on Lower Jurassic and older rocks. It is possible that the Cornubian Ridge and other positive axes within the Western Approaches Basin were also rejuvenated during this time.

Towards the end of the Jurassic a phase of tectonic movements (Late Cimmerian phase) led to uplift, faulting, folding and erosion and the general emergence of the region causing the Late Jurassic sea to withdraw northwards to Lincolnshire and Yorkshire and southwards to Spain. The Early Cretaceous was marked by a return to continental and marginal marine conditions, and thick sequences of fluvial, alluvial, deltaic, lacustrine and coastal swamp deposits were laid down in the isolated depressions of the Channel, Western Approaches and Celtic Sea Basins. Limited sea bottom coring of the Western Approaches has shown the *Lower Cretaceous* to consist dominantly of Wealden continental facies which probably persisted into Albian times. The Lower Cretaceous sediments were deposited on eroded, faulted and tilted Jurassic strata and probably attain thicknesses of up to 150m. Wealden facies rocks occur in Portugal, offshore eastern Canada (Missisanga Formation) and in the Celtic Sea Basins and can be anticipated over the deeper parts of the Western Approaches Basin.[14]

A late Lower Cretaceous phase of crustal disturbance gave rise to a period of strong folding and faulting, the effects of which are well marked in the English Channel. Renewed marine transgression at the end of the Lower Cretaceous produced local overstepping of the older continental beds by younger semi-marine and marine sediments. Continental deposition possibly persisted until late Albian-Cenomanian times. A rapid deepening along the basin axis then resulted in a marine invasion. The *Late Cretaceous* is characterized by a widespread marine transgression which resulted in the submergence of many of the stable platform areas previously established by the Hercynian period of mountain building. This epoch was marked by a rapid deepening of the whole region and the deposition of a fine-grained marine pure limestone known as the *Chalk*. The transgression submerged the upstanding Irish, Western Approaches and London-Brabant Massifs, and the Cornish and Welsh Massifs were reduced to islands rimmed by shoreline sediments which passed rapidly seawards into limestones. A similar relationship occurred along the northern margin of the Brittany Massif.

Since the Late Cretaceous sea was strongly transgressive, strata of this age rest on earlier beds with marked angular unconformity. Chalks comparable to the English Chalk facies predominate, though marginal calcareous sandstones and dolomites occur near the Scilly Isles. In Devon and Dorset the base of the Chalk is sandy and glauconitic which suggests that part of the Cornubian Massif was upstanding at this time. In the south there was progressive transgression onto, and across, the basement rocks of Brittany. Sampling of the offshore seabed outcrops suggests, from faunal evidence, that all the Upper Cretaceous stages may be represented, although the Lower Chalk stages may be condensed or absent towards the east of the basin. The Chalk thickens westwards into the main basin and towards the continental shelf edge, whilst the most complete stratigraphic sequence probably lies along the central axis of the trough. Chalk deposition continued conformably into earliest Tertiary time (Danian).

A pre-Eocene period of earth movements, the *Laramide phase*, gave rise to renewed regional uplift and re-emergence of the platform areas. This was followed by widespread erosion which resulted in the partial removal of the Chalk prior to the deposition of the Eocene sediments. During the Eocene the continental margin was again submerged and the sea periodically invaded the present Western Approaches Basin. As a result shallow marine sands and clays interfinger with fluviatile and deltaic sediments and freshwater limestones. The Start-Cherbourg Ridge may have also been rejuvenated and uplifted during the Laramide earth movements, and thus acted as a barrier between the Western Approaches and Channel Basins from Eocene times onwards. Tertiary sediments cover most of the western part of the basin and thicken towards the edge of the continental shelf where they may attain thicknesses of more than 600m. Uplift again occurred in the Oligocene resulting in a major retreat of the sea, and a period of erosion. In the Western Approaches and Celtic Sea, the coastal zones bordering the uplifted platform areas were the sites of coastal swamps and lake deposits, while within the Western Approaches Basin itself, freshwater limestones were probably deposited. Lignitic clays and sands characterized the isolated large depressions, as in the Bovey Tracey Basin in Devon; the area of the Mochras Borehole at the edge of Cardigan Bay, and other small outcrops in North Devon and Pembrokeshire. Over parts of the Western Approaches Basin Oligocene sediments may also have accumulated in restricted freshwater basins of this type and this, together with a phase of pre-Miocene erosion, may serve to explain the general lack of Oligocene outcrops in the offshore region.

Downwarp of the continental margin occurred in the Miocene with subsidence and marine invasion along the axis of the Western Approaches Basin. The Miocene sea deposited fine-grained sandy and calcareous rocks, containing an abundance of foraminifera ('Globigerina Silts'), with thicknesses of up to 120m. A phase of folding and faulting was accompanied by severe erosion and the removal of Miocene sediments from considerable areas of both the Western Approaches and the Channel. This erosional phase was followed by widespread marine transgression in the Pliocene with the sea invading from the southwest and south, as in Miocene times. Since this period, subsidence has continued along the Western Approaches Basin axis into Recent times. Plio-Pleistocene clays may be 30m thick in places, as indicated by sparker records.

Hydrocarbon Potential

The *source potential* of the pre-Permian rocks of the Western Approaches must be regarded as slight. Many of the rocks are highly tectonized and intruded by late orogenic granites. An exception to this might be Coal Measures in fault-bounded post-orogenic (Stephanian) basins which could act as a source of gas.

The Permo-Triassic sequences, with the exception of the black shales and limestones of the Rhaetic, can also be considered as having no source rock potential. The Jurassic succession, however, has excellent source potential in the thick marine clays and shales which occur, for example, in the Lias and in the Upper Jurassic (Kimmeridge Clay) where present. These marine shales and the coals within the Lower Cretaceous Wealden facies also provide possible source rocks, but the uniform chalk blanket of the Upper Cretaceous has a low source potential. Although the Tertiary succession contains marine shales and carbonates their relatively shallow depth of burial rules them out as important source beds.

Turning to possible *reservoir rocks* in the basin, the pre-Permian is probably too indurated to have much potential. The sandstones of the Permo-Triassic however are thick, and onshore have adequate porosity (for example in the Wytch Farm Oilfield). If structurally juxtaposed against younger source rocks, or where fortuitously occurring above Upper Palaeozoic source rocks, they could provide an adequate hydrocarbon reservoir. The Jurassic, which as noted above has good source potential, also affords interesting reservoir targets. The Jurassic producing horizons of the onshore oilfields at Kimmeridge, Wareham and Wytch Farm have been noted in the previous chapter. In particular sandstones are developed in the Lias (Bridport Sands) onshore which may extend offshore. The original distribution of the Liassic sands was irregular and complex, with the result that they may prove to be difficult and complex exploration targets. Sandstones and carbonates in the Middle Jurassic (Dogger) also have reservoir potential.

In view of the gas production from Wealden and Greensand strata at the Kinsale Head Field in the North Celtic Sea Graben the same interval in the Western Approaches must also be regarded with interest. The Wealden sequence is a variable deltaic sandstone-mudrock sequence, whilst the overlying Lower Cretaceous Greensand comprises a shallow marine coastal sandstone-mudstone sequence. The Lower Cretaceous is generally covered by Chalk and Tertiary rocks and provides interesting drilling targets. The shallow depths of the Chalk and Tertiary sequences, however, render them poor exploration objectives.

The shale and chalk developments in the basin sequence should provide an adequate *seal* for any hydrocarbon traps. The trapping mechanisms are likely to have strong fault influence, although pinchout, fold and unconformity trap possibilities also exist.

Although considerable seismic exploration has been undertaken in the Western Approaches by the oil industry, there has been little deep drilling. The UK extended its offshore designated area in 1971 (Statutory Instrument No. 594) to include areas in the Western Approaches and English Channel. Exploration in the Western Approaches was to some extent delayed by uncertainty regarding the median line between the United Kingdom and France, which was eventually settled by Arbitration. The Arbitration result was handed down in 1977. Further designations then followed over the areas previously in dispute.

In 1977 the UK Ministry of Energy allocated six blocks in the Western Approaches Basin as part of the Fifth Round of licensing, and a further two blocks in the Sixth Round during 1978. 1979 saw the first drilling in the Western Approaches area but BNOC 72/10-1A, 86/18-1 and BP/BNOC 87/12-1A were all plugged and abandoned, and no information has been released. Further blocks have since been awarded in the Seventh Round of UK licensing, and Phillips drilled 73/7-1 during 1981 but this was also dry and abandoned.

The French authorities adopted a different approach to exploration on the western shelf. French companies had begun seismic work in the area in 1961. Conflicting applications had been made to the French Government by a number of companies and groups, dating back to 1967. In 1973 a non-exclusive exploration permit was issued to SNPA over a large part of the Mer Iroise off northwest Brittany. The problem was finally settled by the issue of three large permit blocks in September 1974 to a consortium comprising ELF – SNPA/Total-BP-Shell; the *Iroise* permit northwest of Brittany, *Amor* west of Brest and *Mer Celtique* west to the continental margin. Two deep wells were then drilled on the Armor licence within the French portion of the Western Approaches basin. The Lizzen-1 well 110 km northwest of Brest drilled in 1975 to more than 4,000m is rumoured to have lacked good reservoirs and to have penetrated a thick mudrock – marl section. A second well, Lenkett-1, was also abandoned as a dry hole (total depth 2,762m) although oil shows were encountered at about 1,780m. Brezell-1, the third well, bottomed at 3,337m and was plugged and abandoned as a dry hole.

After this disappointing start to exploration in the region the companies went through a period of consolidation during which a considerable amount of new reflection seismic data was acquired. In 1978 two further unsuccessful wells were drilled off the coast of Brittany; Levneg-1 (3,255m) west of Brest in the Mer Celtique permit and Penma-1 (1,640m) in the new Loire Maritime permit south of Brest. 1979 saw two further dry holes, Glazenne-1 (3,104m: 7°53′W 48°05N) and Yarvor-1 (3,453m: 6°22′W 48°14′N).

Further prospecting authorizations were granted in the Western Approaches – Channel areas in 1980 to SNEA (P) and Coperex, and seismic programmes were carried out. Two wells were drilled and abandoned in 1980, Kulzenn-1 (3,165m: 4°07′W 49°25′N) and Travank-1 (1,645m: 5°24′W 48°33′N) north and west of Brest respectively. Two further wildcats were scheduled by SNEA in 1981 on the Atlantic seaboard—Rea Gwen-1 (6°52′W 48°35′N) which was dry and abandoned at 2,525m T.D., and Galizenn-1 (7°39′W, 48°21′N; T.D. 2,526m) which was also dry and abandoned.

After drilling ten dry holes (four in Armor, three in Mer d'Iroise and three in Mer Celtique) and spending 600 million francs, the consortium is reported to be limiting work in 1982 to seismic programmes (2,000 km). The aim will be to decide which areas to retain when the permits come up for renewal in May 1983.

References

1. KING, W. B. R. 1954. The geological history of the English Channel *Q. J. geol. Soc. London* **110**, 77–101.
2. CURRY, D., HAMILTON, D. and SMITH, A. J. 1970. Geological and shallow geophysical investigations in the Western Approaches to the English Channel. *Inst. geol. Sci. Rep.* 70/3, 12pp.
3. CURRY, D., HAMILTON, D. and SMITH, A. J. 1971. Geological evolution of the western English Channel basin and its relation to the nearby continental margin: *in* Delany F. M. (*Ed.*), The geology of the east Atlantic continental margin. Pt 2: Europe. *Inst. geol. Sci. Rep.* 70/14, 129–142.
4. DAY, A. B., HILL, M. N., LAUGHTON, A. S. and SWALLOW, J. C. 1956. Seismic prospecting in the Western Approaches of the English Channel. *Q. J. geol. Soc. London* **112**, 15–44.
5. BACON, M. 1975. A gravity survey of the western English Channel between Lyme Bay and St. Brieuc Bay. *Philos. Trans. R. Soc. London* **A 279**, 69–78.
6. HILL, M. N. and VINE, F. J. 1965. A preliminary magnetic survey of the Western Approaches to the English Channel. *Q. J. geol. Soc. London* **121**, 463–475.

7. AVEDIK, F. 1975. Seismic refraction survey in the Western Approaches to the English Channel: preliminary results. *Philos. Trans. R. Soc. London* **A 279**, 29–39.
8. AVEDIK, F. 1975. The seismic structure of the Western Approaches and the South Armorican continental shelf and its geological interpretation: *in* Woodland, A. W. *(Ed.)*, *Petroleum and the Continental Shelf of North West Europe. 1 Geology*. Applied Science Publishers, London, 29–43.
9. SMITH, A. J. and CURRY, D. 1975. Structure and geological evolution of the English Channel. *Philos. Trans. R. Soc. London* **A 279**, 3–20.
10. ANDREIEFF, P., BOUYESSE, P., CURRY, D., FLETCHER, B. N. HAMILTON, D., MONCIARDINI, C. and SMITH, A. J. 1975. The stratigraphy of post-Palaeozoic sequences in part of the Western Channel. *Philos. Trans. R. Soc. London* **A 279**, 79–97.
11. BUREAU DE RECHERCHES GÉOLOGIQUES ET MINIÈRES 1974. Geological map of the Continental Margin: English Channel. Scale 1:1,000,000, with *Explanatory Notes*.
12. HAMILTON, D. 1979. The geology of the English Channel, South Celtic Sea and Continental Margin, South Western Approaches: *in* Banner, F. T., Collins, M. B and Massie, K. S. *(Eds)*, *The North-West European Shelf Seas 1. Geology and Sedimentology* Elsevier, Amsterdam–Oxford–New York, 61–87.
13. WARRINGTON, G. *et al.* 1980. A correlation of Triassic rocks in the British Isles. *Geological Soc. London, Special Report No. 13.* 78pp.
14. ALLEN, P. 1981. Pursuit of Wealden models. *J. geol. Soc. London* **138**, 375–405.
15. NAYLOR, D. and MOUNTENEY, S. N. 1975. *Geology of the North-West European Continental Shelf. Vol. 1* Graham, Trotman and Dudley, London, 162pp.
16. NOROIL, January 1982, p. 64.

Chapter 4

The Celtic Sea Basins

Introduction

The continental shelf between Ireland and southwest England is covered by the relatively shallow water of the Celtic Sea. As a general rule water depths do not exceed 200 m. This extensive shelf area is underlain by a pair of east-northeast-trending, parallel, fault-bounded, deep and generally complex sedimentary troughs known as the *North* and *South Celtic Sea Grabens*. These are bounded to the north by Devonian and Carboniferous metasediments which crop out on the Irish mainland and which are covered by a veneer of Upper Cretaceous, Tertiary and Quaternary sediments on the basement platform area lying immediately south of the mainland. The southern boundary of the grabens is formed by the Hercynian *Cornubian Massif* which is exposed in the Devon and Cornwall peninsula and which extends southwestwards as an elongate shallow subsurface platform. This platform is intruded by a number of Hercynian granites, notably the Haig Fras Granite, while a number of subsurface granites are indicated by negative gravity anomalies. The Western Approaches-English Channel Basin, paralleling the east-northeast trend of the Celtic Sea Grabens, lies to the south of the Cornubian platform. The North Celtic Sea Graben is separated from the narrower South Graben by the *Pembrokeshire Ridge* and its southwestwards continuation known as the *Labadie Bank Basement High* (Fig. 4.1). The ridge system lies close to a median line between Ireland and Britain. It forms a shallow elongate discontinuous subsurface platform cored by Palaeozoic metasediments and appears to be covered by a condensed sequence of Mesozoic sediments. This platform is likely to have acted as an intermittently-active partial barrier between the two grabens during most of the Mesozoic Era.

The small and narrow Fastnet Basin lies immediately westwards of the North Graben and both troughs have broadly comparable geological sequences, perhaps indicating sedimentary interconnection during much of their evolution. The North Graben runs east-northeastwards, parallel to the southern Irish coastline as far as Carnsore Point. It then turns northeastwards into the southern Irish Sea where it terminates in a number of complex small structural units. The most notable of these are shown on Fig. 4.1, and they include the *St. George's Channel* and *Cardigan Bay Basins* which are separated by a low-relief southwards extension of *St. Tudwal's Arch*, and the *Central Irish Sea Basin* which is separated from the St. George's Channel by the *Irish Sea Geanticline.*

The South Celtic Sea Graben parallels the North Graben northeastwards almost as far as Pembrokeshire, where the Pembrokeshire Ridge widens appreciably. The South Graben then narrows substantially and extends eastwards through the *Bristol Channel* and inland to link up with the north-south trending Cheshire and Worcester Grabens and the major Wessex Basin.

This chapter describes the geology of most of the main basins in the general Celtic Sea – southern Irish Sea area. It includes most notably the North Celtic Sea Graben and its eastward extension into the St. George's Channel and Cardigan Bay Basins, and the South Celtic Sea Graben with its eastward extension into the Bristol Channel Basin. In view of the large volume of geological and geophysical data available, the small Fastnet Basin (Fig. 4.1) is not discussed here but is held over and described separately in the following chapter.

The area now covered by the central part of the Celtic Sea probably lay in a marine environment for much of Lower through Middle Carboniferous times.[1] Shallower shelf or deltaic conditions existed during much of this time period to the west in the Fastnet Basin area and to the northeast in the St. George's Channel and south Irish Sea areas. However, it is likely that even in Upper Palaeozoic times the proto-Celtic Sea Grabens with their pronounced caledonoid trend were in existence, although subsidence and basin infill in Palaeozoic times were likely to have been minimal. This early sedimentation terminated with the onset of the Hercynian orogenic episode in Late Carboniferous – Early Permian times during which the existing sediments which now comprise the economic basement for the area were tectonized, folded and indurated. This period of compression was followed by episodic extension which led to the active subsidence of the grabens and basins in the area. The sedimentary basins of Newfoundland and Eastern Canada (Grand Banks and Scotian Shelf) lay adjacent to the infant Celtic Sea Grabens and continued the caledonoid east-northeast trend westwards (Fig. 1.3). However, it is unlikely that there was any actual sediment pathway between the Irish and Canadian basins at this early stage in their evolution.

The initial Mesozoic sedimentation was Triassic in age and was of continental fluviatile origin with the surrounding uplifted Hercynian landmasses providing the sedimentary detritus. Deposition in arid conditions with consequent evaporation in inland lakes and seas gave rise to localized thick sequences of salt during Upper Triassic times.

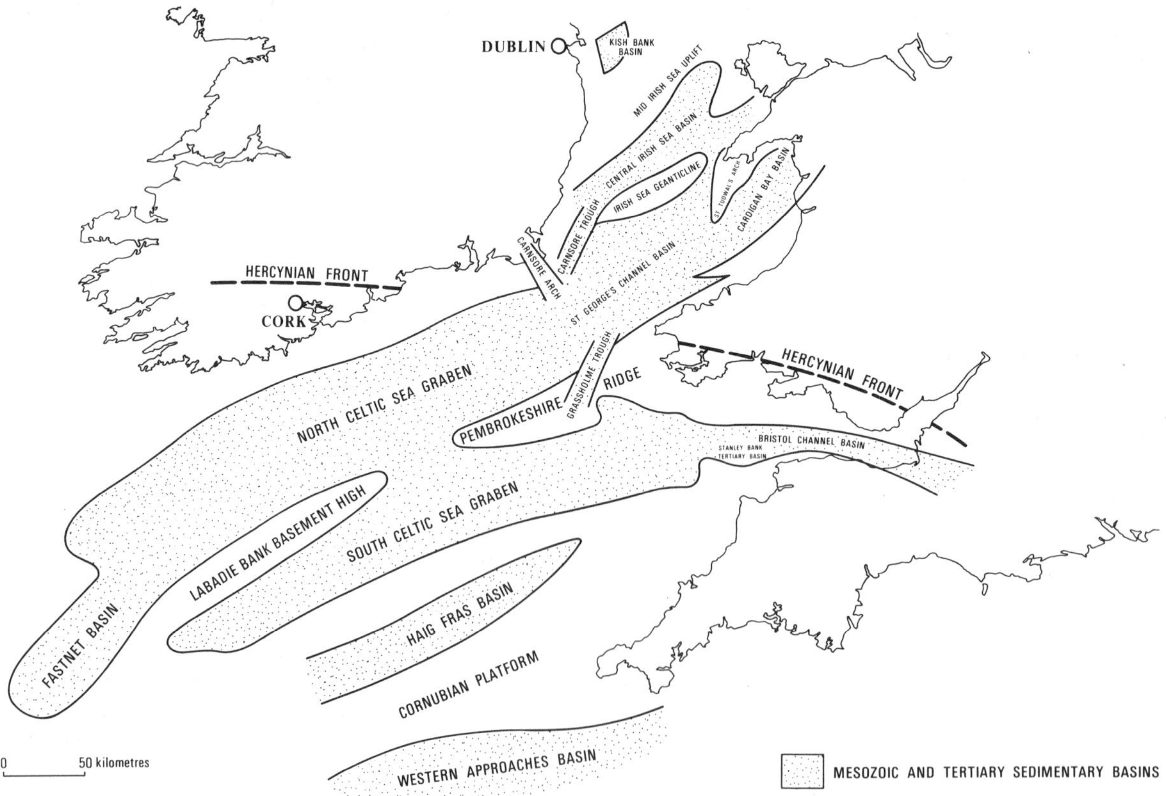

Figure 4.1. Location map showing Celtic Sea Grabens and associated basins and main structural elements.

In Late Triassic – Early Jurassic times a widespread and rapid marine transgression inundated both grabens and also covered the low-lying areas which lay between Britain and North America giving rise to widespread shallow marine shelf conditions. Depositional environments deepened slightly throughout the Early and Middle Jurassic while more restricted shelf conditions again returned in Upper Jurassic times.

A major period of uplift, fault block movement and igneous intrusion occurred at the end of Middle Jurassic time. This was probably associated with the initial stages of the formation of the North Atlantic Ocean when eastern North America began to separate from the western margin of the Iberian Peninsula.

The Jurassic-Cretaceous boundary is often marked by a strong unconformity which reflects the widespread Late Cimmerian phase of activity recognized throughout much of North West Europe.[1] The continental influence on the Lower Cretaceous sediments was terminated in Late Cretaceous time by a major marine transgression, from which only the higher ground of Ireland, Wales and Cornwall remained emergent. Its commencement is approximately synchronous with the initiation of the separation of the Irish and the North American basins. Shallow shelf seas deposited a thick sequence of Chalk over the Mesozoic basinal regions and a thinner, sometimes discontinuous, layer across the submerged platforms dividing the North and South Celtic Sea Grabens and the Western Approaches Basin (see Chapter 3) and also across the southern part of the Irish mainland platform area. Much of this thin Chalk veneer was removed from the platform areas during subsequent erosion in Tertiary and Quaternary times. However, an isolated remnant of Upper Cretaceous Chalk occurs onshore in southern Ireland and gives an indication of the large areal extent of the shallow Chalk seas.[2]

The top of the Upper Cretaceous is often marked by a major unconformity, the formation of which was approximately coincident with the commencement of sea floor spreading along the Reykjanes Ridge. The overlying Tertiary is represented by a sequence of probably non-marine fluvial clastics and shallow marine sediments followed by deeper marine fine grained clastics. The Palaeocene – Eocene boundary approximately coincides with a major Alpine unconformity which reflects the collision between the African and Eurasian continental plates. However, as with underlying Upper Cretaceous, it is likely that Tertiary sedimentation may have originally extended over a much wider geographical area. At a later stage glacial erosion and deposition resulted in the distribution of a generally thin veneer of Quaternary sediments throughout much of the Celtic Sea area.

Geophysical Data

Although published geophysical data, and in particular deep seismic reflection information, are fairly limited, a large quantity of geophysical information has been acquired by oil companies, geophysical contractors and academic institutions. For instance, since geophysical surveying began in the Irish sector of the Celtic Sea area in 1969 a total in excess of 20,000 km of seismic reflection data has been shot. A substantial amount of this is now available for purchase by oil exploration companies from the Department of Industry and Energy in Dublin. In addition to this, various gravity and magnetic surveys have been carried out and the results have been published by a number of scientific institutions in the United Kingdom.[3–5]

Some very limited seismic refraction work[6,7] in the Celtic Sea area indicates that the general area is underlain by continental crust which is something of the order of 28 km thick. Very little stretching or resultant thinning of the crust is therefore suggested. This work points to the existence of a horizontal Moho in the area, lying at a depth of between 27 and 31 km.

The gravity work[5] reveals a broad pattern of gravity lows corresponding to the depocentres in the Celtic Sea Grabens. The well-defined gravity highs reflect the uplifted platforms, largely composed of economic basement, which border the basins and which also occur along the Pembrokeshire Ridge. A series of gravity lows are seen over a number of areas within the Pembrokeshire Ridge and the southern margin of the Celtic Sea Grabens. These probably reflect shallow Hercynian granites similar to those of the onshore Cornubian Massif. Gravity profiling reveals a prominent east-west Hercynian grain extending throughout the South Celtic Sea Graben and Bristol Channel areas. A northeast-southwest caledonoid basement trend occurs in the Cardigan Bay and St. George's Channel Basins. Followed southwestwards along strike it swings round into the Hercynian trend seen in the North Celtic Sea Graben.

The seismic data for the Celtic Sea between southern Ireland and the Cornubian Massif show that this marine area is underlain by two deep asymmetrical grabenal basins which are infilled by some 3–9 km of sediment.[8] The basins are separated by the Pembrokeshire Ridge-Labadie Bank Basement High system which extends discontinuously southwestwards from St. David's Head in Pembrokeshire (Fig. 4.1). Extensive down-to-the-basin faulting, parallel to the basin margins, occurs in both basins while an orthogonal northwest-southeast direction is also present but is far less prominent. The orientations, shapes and sizes of the various basins in the area appear to be governed by an interplay of the predominant basement trends. A simple east-northeast trend is evident in the South Graben and the western part of the North Graben, while moving northeastwards the pattern becomes more complex and the area is broken up into a number of smaller basins and sub-basins separated from each other by basement highs.

Deep seismic reflection information reveals the general presence within the Grabens of a number of low-velocity layered units resting unconformably on a high velocity basement. In the St. George's Channel and Cardigan Bay areas the uppermost layer probably corresponds to unconsolidated Quaternary material. This thins southwestwards and is only patchily developed over the North and South Celtic Sea Grabens. The underlying layers of this seismic stratigraphy may usually be identified as follows. A Tertiary to Upper Cretaceous layer extends throughout most of the Celtic Sea area with the exception of the southern Irish Sea, Cardigan Bay and Bristol Channel areas. Beneath this lies a thick Lower Cretaceous layer which is also present in the North and South Grabens, and in the St. George's Channel Basin. A variable, but normally thick, Jurassic layer overlying a Permo-Triassic sequence occurs throughout most of the Celtic Sea basins. However, due to the presence of a highly reflective near-surface thick Senonian – Turonian chalk sequence over much of the area (Fig. 4.2) the general quality of pre-1981 seismic data at pre-Cretaceous Chalk levels tends to be rather poor and therefore definition of structures and sediment thicknesses on seismic grounds in these areas is uncertain. Recent advances in seismic acquisition techniques using high pressure airgun sources and long geophone cables appear to have substantially improved the data quality in the area. An enhancement of the present geological understanding of the area at the deeper Jurassic and even Permo-Triassic levels is likely to follow this improvement in the near future.

Stratigraphy

At the time of writing (early 1982) a large number of exploration wells have been drilled in the general Celtic Sea area. Twenty eight exploration wells were drilled in the North Celtic Sea Graben, mostly by Marathon and Esso, with one each by Elf Aquitaine and British Petroleum. Very little geological information has been released by the companies involved in the exploration. Eight wells (three in Irish waters and five in UK designated territory) have been drilled in the South Celtic Sea Graben and published information on the resultant geological data is even more scant than for the North Graben. Due to two recent publications the geology of the St. George's Channel Basin,[9] where six offshore wells have been drilled to date, and the Bristol Channel[10] where three wells were drilled, is relatively well known. One well was also drilled in the Carnsore Trough (Fig. 4.1). The evolution of the adjacent Fastnet Basin (Chapter 5) is also well documented.[11] The geology of the Celtic Sea area as a whole can therefore be compiled with a high degree of confidence, using the existing published well data from the general area, together with available seismic information and the results of fairly extensive sea floor sampling carried out by the University of Aberystwyth.[12,13]

The high velocity basement which underlies the Grabens and which forms the dividing platform (Pembrokeshire Ridge) between them is probably of Upper Palaeozoic age. In the southern onshore Ireland the equivalent basement comprises folded and slightly metamorphosed Devonian and Lower Carboniferous redbeds, carbonates and fine-grained clastics with small pockets of Upper Carboniferous strata. In the Fastnet Basin the basement comprises Devonian clastics and Lower Carboniferous shelf limestones. Within the Celtic Sea Grabens at least three wells (Marathon 48/30-1, Marathon 58/3-1 and Esso/Marathon 56/20-1) bottomed in Carboniferous shales[14,15] thereby indicating the presence in the area of an Upper Palaeozoic marine, non-carbonate environment. Palynological studies on a shale sample obtained by a University College, Aberystwyth research cruise on the shelf north of the North Graben indicate a mid-Tournaisian age[13] while a Tournaisian age was also obtained from the Esso-

Marathon 48/30-1 well.[15] The lithology and age of the sample indicate a southwards extension of the relatively shallow marine conditions during the Tournaisian which obtained on the mainland.

Much of the southern part of the North Sea is known to be underlain by Westphalian and Stephanian Coal Measures.[1] Coal Measures are extensively developed along the southern margin of the Palaeozoic St. George's Land. Westphalian 'C' and 'D' Coal Measures are also recorded from the Kish Bank Basin east of Dublin (see Chapter 13) and in the Texaco/ HGB 103/2-1 well in the UK sector of the St. George's Channel Basin.[9] In addition they are preserved extensively onshore in South Wales. It might therefore be expected that they occur towards the extreme eastern part of the North Celtic Sea Graben and the southern part of the Irish Sea. A cyclic sequence of shales, sandstones and thin coal beds, deposited in a brackish swamp estuarine environment, might be anticipated.

During the Hercynian (Armorican) phase of deformation which succeeded deposition of the Upper Palaeozoic sediments, the rocks were indurated, heated and deformed. This deformation increased in intensity southwards ac-

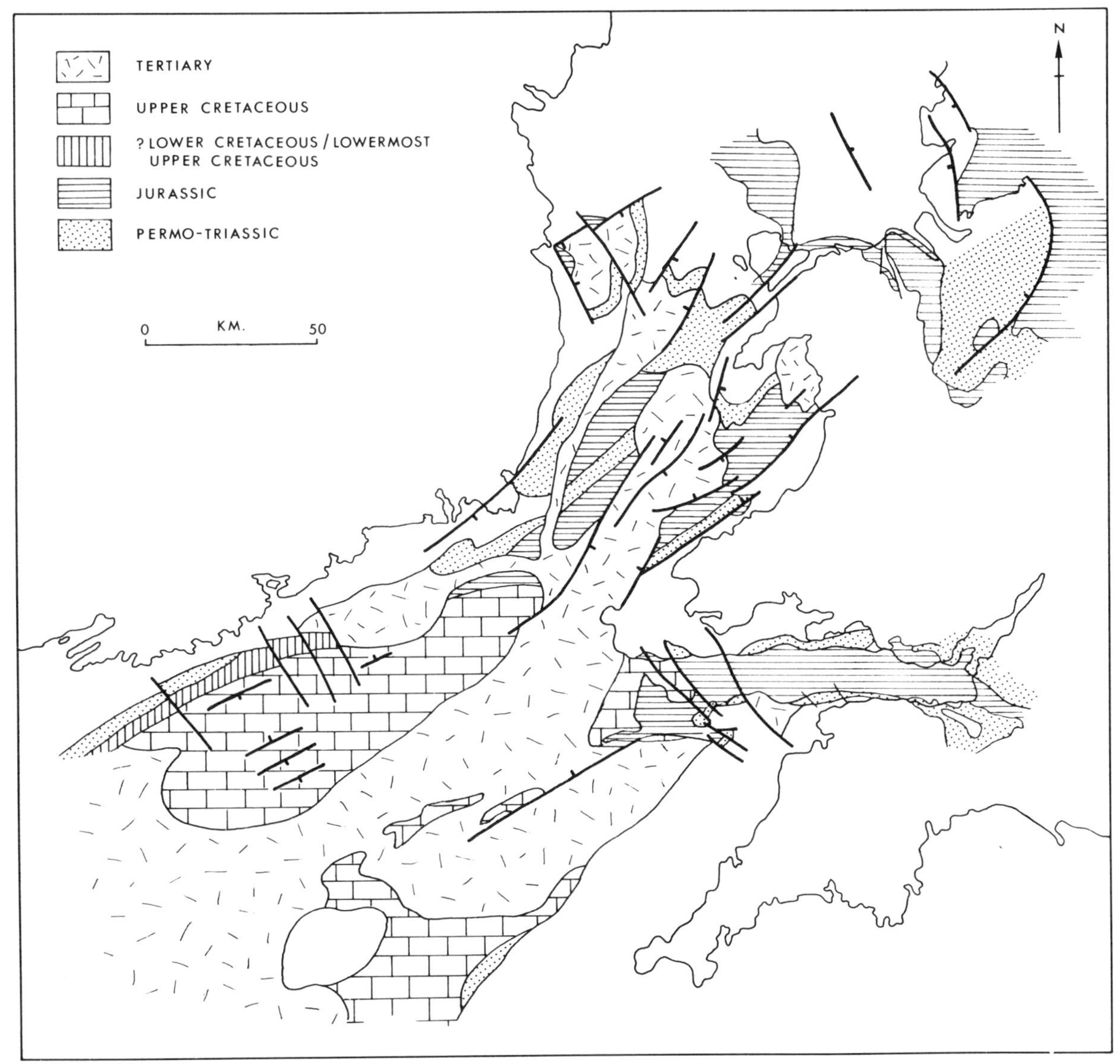

Figure 4.2. Map showing the subcrop of pre-Quaternary strata in the Celtic Sea area. (Modified largely from Blundell (1979).[5])

ross the Cornubian Massif to the Brittany coast. Here, towards the core of the Hercynian mountain chain, the Upper Palaeozoic rocks were progressively more strongly deformed and were intruded by a number of granite bodies, most notably the Cornish, Brittany, Haig Fras and Scilly Isles granites.

Following this period of deformation, uplift and erosion the relaxation of compressive stresses led to fault-controlled downwarping in the Celtic Sea area, with continental Triassic sediments deposited unconformably on the Palaeozoic basement. Outcrops of Triassic rocks now fringe the northern and eastern margins of the North Celtic Sea Graben and the eastern margin of the South Graben (Fig. 4.2). They are widespread on the seafloor of the St. George's Channel and Central Irish Sea area. They crop out extensively onshore in South Wales, Gloucestershire and Somerset, and also on the Devon/Cornwall peninsula where upwards of 2,500m of continental red siltstones and sandstones overlie the basement strata.

The North and South Celtic Sea Grabens contain a relatively thick Triassic sequence. Somewhat in excess of 600 m of Triassic sediments were recorded in the Esso/Marathon 56/20-1 well.[14] In this well, which lies towards the southern flank of the North Graben, the Triassic succession is of continental origin and comprises a lower sandstone-dominated section, a middle thin calcareous section and a thick (450m) Upper Triassic sequence of siltstones and shales with sandstones which decrease in number upwards in the section (Fig. 4.3). Evaporites, especially anhydrite, are common in the upper portion of this (probably Keuper) fine-grained section while some volcanic rocks are also recorded. A similar well-developed Triassic redbed sequence, topped by evaporites, has been encountered further east in the North Graben in wells drilled by Marathon.[8] Sea bed samples of Triassic strata taken from the northern margin of the North Graben[13] indicate a similar continental clastic succession.

In the Cardigan Bay area the Mochras borehole penetrated approximately 50m of Late Triassic to Rhaetic sediments while a number of offshore wells in the St. George's Channel Basin encountered thicknesses of up to 2,000m of Triassic sediments (Fig. 4.4) comprising sandstones overlain by anhydritic and halitic mudstones.[9] Evidence for salt diapirism in this area is also seen on seismic profiles. Sea bed samples from St. Tudwal's Arch yielded presumed Permo-Triassic red evaporitic marls and mudstones.

Eastwards of the South Celtic Sea Graben Permo-Triassic sediments occur in the Bristol Channel Basin and the onshore Quantock-Mendip Basin in England.[10] The Bristol Channel Basin contains up to 1,200m of Permo-Triassic sediments (Fig. 4.5). This represents a narrow and eastwards-shallowing sedimentary trough between the large South Celtic Sea Graben to the west and the Quantock – Mendip Basin to the east, where in the Puriton Borehole of Somerset over 350m of Upper Triassic marls and evaporites overlie 300m of Lower Triassic and Permian coarse conglomerates and sandstones. Sea bed sampling shows that the Triassic of the Bristol Channel is similar to the Keuper of west Somerset. It consists of red silty clays and fine-grained continental sandstones overlain by a succession of Keuper marls and Rhaetic shales and sandstones. Along the northern coast of the Bristol Channel, in the vicinity of Cardiff, Triassic rocks rest unconformably on deformed Devonian and Carboniferous metasediments.[16] Permian and Lower Triassic rocks are absent and the oldest Mesozoic rocks are Keuper in age. These grade laterally from coarse lacustrine scree deposits developed on a number of wave-cut platforms into characteristic Keuper Marls containing diagenetic nodular and locally bedded evaporites which reflect periods of evaporation in the shallow Keuper basin. Evidence of Keuper aeolian deposition into restricted super-saline waterbodies is also seen in parts of the basin. An eastward-thinning Keuper salt section occurring within the clastic sequence is deduced from the offshore well information in the basin.[10] In this area a series of transitional beds marks the gradual change from the continental red bed facies of the Triassic into the marine facies of the overlying Jurassic. They comprise a fine-grained Rhaetic sequence of dark shales with occasional thin limestones which underlie the Liassic section. In the west and east of the basin the transitional beds are of Rhaetian age, while in the centre (Shell 103/18-1) they are of Hettagian age.[10]

Jurassic rocks do not crop out extensively on the sea bed of the Celtic Sea. This is probably due to the extensive blanketing by the later Cretaceous and Tertiary sediments. They do, however, occur on the sea floor in the St. George's Channel, Cardigan Bay and extensively in the Bristol Channel Basins (Fig. 4.2).

A thick sequence of Jurassic sedimentary rocks occurs in both the North and South Celtic Sea Grabens. A comparison between the patterns of Liassic sedimentation in the Fastnet

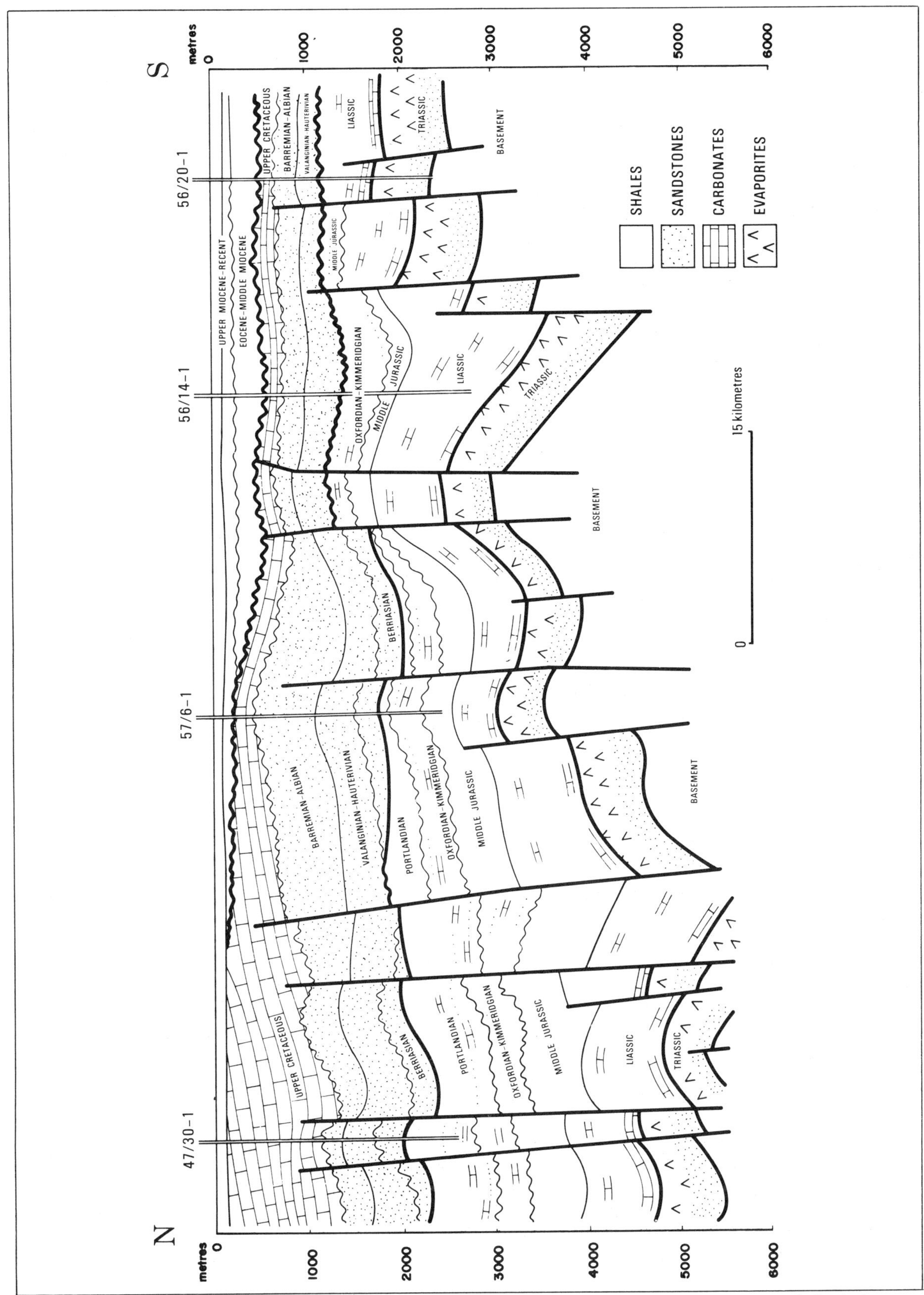

Figure 4.3. Approximate north-south cross-section through the North Celtic Sea Graben. (After Colin, Lehman and Morgan (1981).[14])

Basin to the west, and the Bristol Channel, St. George's Channel and Cardigan Bay Basins to the east, together with published information from a number of Esso/Marathon wells (notably 56/20-1 and 56/14-1) suggests that the Liassic in the Celtic Sea Grabens follows upon the Triassic with a marine transgression of approximately Rhaetian to Hettangian age, and is represented by marine marly shales with a basal shelf limestone. An eastward decrease in the thickness of the limestones could be expected with stable shelf conditions persisting longer in the west of the Celtic Sea area. The Hettangian section is probably overlain by Sinemurian to Aalenian marly shales which reflect a slight and gradual deepening of depositional environment.

The Middle Jurassic (pre-Oxfordian) succession appears in general to conformably overlie the Liassic with the continuation of marine shale deposition. Shallow marine shelf limestones and minor sandstones may also be expected.[1]

Middle and Upper Jurassic strata occur in both the North and South Celtic Sea Grabens and are generally separated by an unconformity. In the centre of the North Graben the Middle Jurassic is more than 750m thick while the Upper Jurassic is in excess of 1,200m.[14] The Middle and Upper Jurassic sections display a clear thinning and wedging out southwards towards the southern basin margin of the North Graben (Fig. 4.3).

Tectonic movements in Middle Jurassic times, corresponding to the Main Cimmerian Phase widely recognized throughout Northwest Europe[1] probably led to the uplift of the graben margins and to differential fault block movement within the basins. As a result, the succeeding Oxfordian, Kimmeridgian and Portlandian successions are likely to show rapid and extensive facies and thickness variations throughout the area. Appreciable quantities of non-marine (probably fluvial, alluvial and deltaic), brackish and shallow shelf sediments occur.[1,14] The main lithologies in the Upper Jurassic section are shales and sandstones with Kimmeridgian and Portlandian limestones commonly recorded.[14] A number of unconformities are recorded in the Upper Jurassic section (Fig. 4.3), especially in the western part of the North Graben. These are also probably a reflection of differential fault block movement. The Portlandian sediments are generally succeeded by Lower Purbeckian shales in the centre of the North Graben,[17] while towards the margins of the basin the Jurassic-Cretaceous boundary tends to be unconformable (Fig. 4.3) with the development of the dominantly non-marine to brackish Purbecko-Wealden facies. This sequence represents the rather poorly defined and apparently complex junction between the Upper Jurassic and Lower Cretaceous sections in the Celtic Sea. This in turn is overlain by the ubiquitous non-marine clastic sequence of the classic Wealden facies.

The St. George's Channel Basin, and particularly the eastern part, contains substantial thicknesses of Jurassic sediments. The upper Jurassic (Oxfordian to Purbeckian) of lagoonal to marginal marine dominantly argillaceous lithofacies exceeds 1,200m in places while the Middle Jurassic (Bajocian to Callovian) is sometimes in excess of 1,000m. In common with the adjacent and connected Cardigan Bay Basin, the Liassic (pre-Pliensbachian) section in the St. George's Channel Basin is about 400m thick (Fig. 4.4).

The Mochras borehole at the eastern end of the Cardigan Bay Basin penetrated 1,300m of Liassic (Hettangian to Toarcian) shallow shelf open marine mudstones, siltstones and limestones. On St. Tudwal's Arch up to 800m of Middle Jurassic (Bajocian and Bathonian) shales and shallow water sandstones and limestones overlie approximately 400m of Liassic sediments. Middle Jurassic strata crop out on the seabed in a number of parts of Cardigan Bay. A substantial thickness of Lower and Middle Jurassic is therefore indicated for the basin. The St. Tudwal's Arch was positive during the Middle Jurassic, resulting in a thin and condensed sedimentary sequence over the Arch, while minor unconformities representative of the Mid Cimmerian Phase[1] are recorded in top Liassic strata, and minor warping and erosion is also seen in the Cardigan Bay Basin.

In the Bristol Channel Basin, Lower and Middle Jurassic strata comprise the dominant part of the Jurassic sequence and are, in broad terms, lithologically similar to those of the Celtic Sea Grabens and Fastnet Basin to the west (Fig. 4.5). They consist predominantly of marine calcareous claystones and marls with thin limestone intercalations. Limestones appear to be best developed in the Hettangian and Sinemurian parts of the section. In the west of the basin only the Lower Jurassic is preserved, with thicknesses of the order of 600m being recorded. The dominantly argillaceous Middle Jurassic occupies the central portion of the basin where it conformably overlies the Liassic. In this area 100m of Upper Jurassic (Kim-

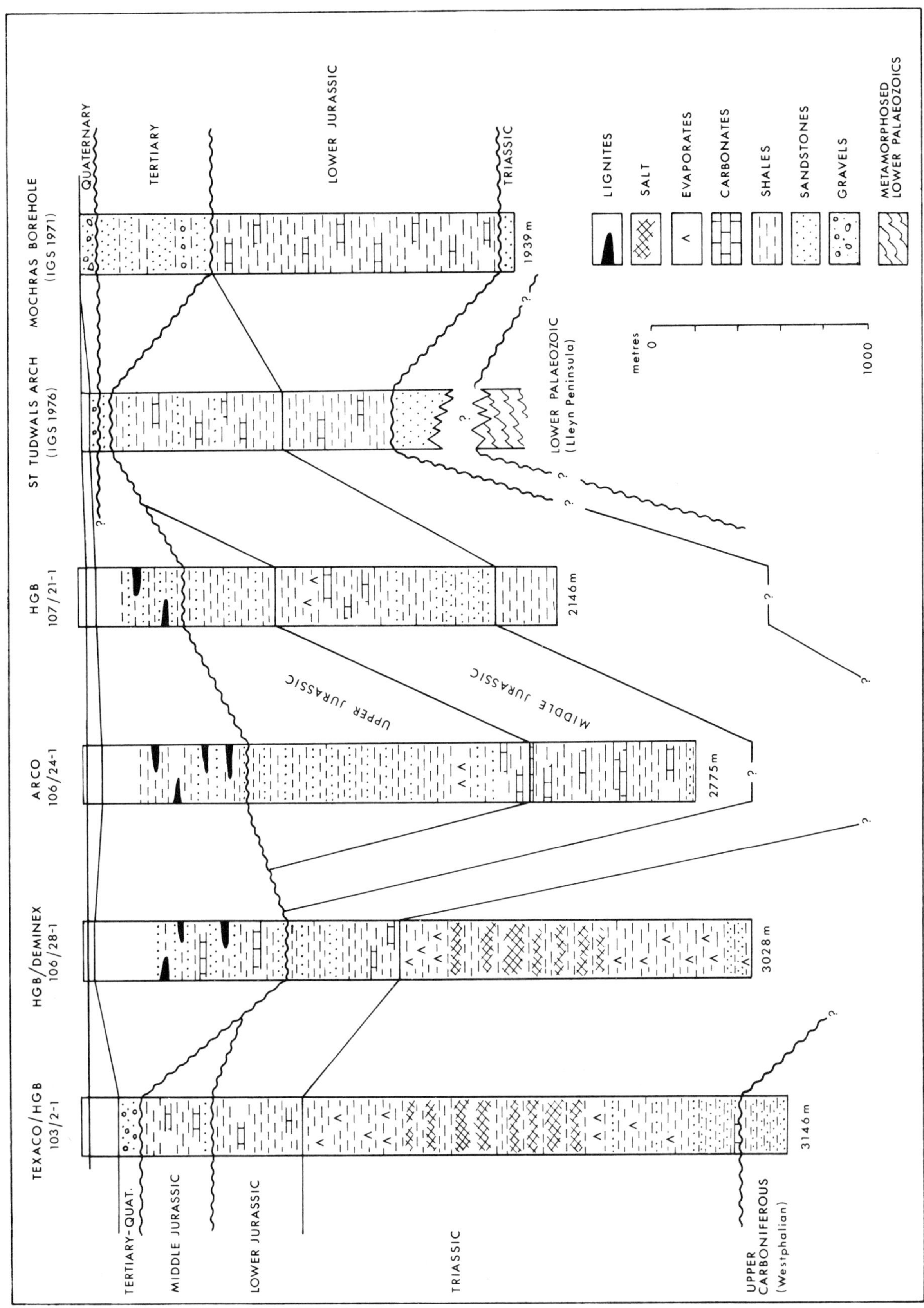

Figure 4.4. Correlation diagram of wells and surface sections in the St. George's Channel-Cardigan Bay area, (after Barr, Colter and Young (1981)[8]).

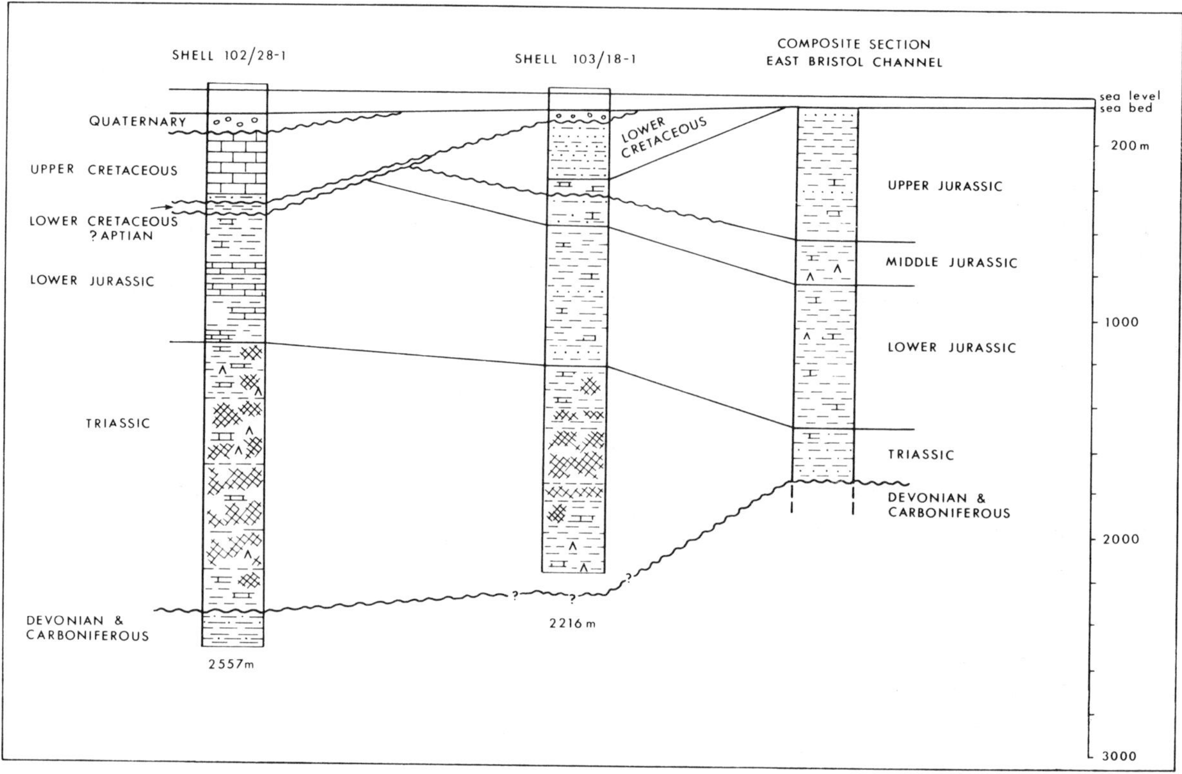

Figure 4.5. Correlation diagram of wells and surface sections in the Bristol Channel area, (after Kamerling (1979)[9]).

meridgian–Portlandian) regressive inner neritic sediments disconformably overlie Middle Jurassic (Callovian) claystones. In the centre of the basin a number of thin Pliensbachian sandstones are developed, perhaps somewhat analogous to the Sinemurian delta fringe deposits of the western part of the Fastnet Basin (see Chapter 5). In the eastern part of the Bristol Channel there appears to be a continuous uninterrupted Jurassic succession some 1,500m thick, which possibly spans the Jurassic-Cretaceous boundary.

Cretaceous rocks have a widespread occurrence on the sea floor of the Celtic Sea Grabens (Fig. 4.2). Further east they crop out towards the western end of the Bristol Channel Basin but are absent from the floor of the St. George's Channel and Cardigan Bay Basins. Drilling has confirmed the absence of Lower and Upper Cretaceous sediments in both these areas. In the western and possibly central parts of the Bristol Channel Basin, Jurassic strata are locally unconformably overlain by an eastwards-thinning sequence of Cretaceous sediments. The Lower Cretaceous comprises a maximum of 300m of marginal marine, probably Aptian, clastics. These are unconformably overlain by a minimum of 1,000m of Upper Cretaceous (principally Cenomanian to Campanian) chalk. Both Lower and Upper Cretaceous strata are absent from the east of the basin.

Although only limited areas of possible Lower Cretaceous rocks crop out on the sea floor close to the northern margin of the North Celtic Sea Graben, such sediments have been repeatedly encountered during drilling in the Celtic Sea. In the Fastnet Basin to the west, uplift, often inversion-related and possibly connected with thermal arching along the continental margin to the west,[18] together with extensive erosion in Late Jurassic-Early Cretaceous times, provided a sediment source for the overlying continental Wealden sediments deposited in the North Graben. A similar break occurs in the Bristol Channel Basin, but in both the South, and especially the North, Celtic Sea Grabens this unconformity is only developed towards the margins of the grabens. In the centre of these basins sedimentation was continuous from the Upper Jurassic into the Lower Cretaceous. This variation in the nature of the Jurassic-Cretaceous boundary is illustrated in Fig. 4.3. In the Esso/Marathon 47/30-1 and 57/6-1 wells, drilled in the central part of the North Graben, no distinct regional break occurs between the uppermost Jurassic and lowermost Cretaceous. A Berriasian sequence of non-marine deltaic clastics conformably overlies Upper Jurassic Portlandian deposits. However, to the south in the Esso/Marathon 56/14-1 well the Berriasian is apparently missing and a well-defined Late Cimmerian Unconformity is seen with the Lower Cretaceous (Valanginian-Hauterivian) sediments overstepping the alluvial and fluvial sediments of the Oxfordian-Kimmeridgian. Southwards towards the Pembrokeshire Ridge the Lower Cretaceous rests unconformably on Lower Jurassics.

The Cretaceous in the Celtic Sea Grabens is broadly subdivided into a continental sandy (Wealden) section overlain by a marine shaly, sandy and calcareous sequence. The earliest Cretaceous sedimentation in the North Graben was Berriasian in age and probably comprised a brackish to freshwater, generally alluvial, sequence. This was followed by Valanginian to Hauterivian continental deposits, probably in the form of fluvial sandstones and alluvial shales, with the main source area lying approximately to the west.[11,18] In Hauterivian times faulting, foundering and ultimate collapse of the updomed continental margin to the west[18] is likely to have beheaded this source area and resulted in the deposition of only muddy alluvial deposits in this western part of the basin. The upstanding Irish and Welsh Massifs lying to the north and east are presumed to have continued pouring coarser sediments into the eastern part of the area. By Barremian times the first pulses of the Cretaceous marine transgression began to occur in the west of the graben while continental fluvial sedimentation continued in the east.[14] In the centre of the graben a complex pattern of fluvial, alluvial and deltaic sedimentation is likely to have occurred during Barremian times with the deposition of sandstones, siltstones, mudstones and minor coals. This is illustrated in the vicinity of the Kinsale Head Gas Field[17] where the sandstones display typical cyclical and coarsening-upward fluvial characteristics. Individual units are typically 3–10m thick but may not be laterally very continuous. Major distributory channels in excess of 25m rarely occur. The argillaceous sediments in the area reflect levee and overbank sequences while interdistributory swamp and marsh environments are represented by the coals, palaeosols, carbonaceous shales and pyritiferous claystones.

By Aptian times the marine transgression had become fairly well established in the western half of the North Graben and was rapidly moving eastwards, while non-marine deposition probably continued in the east of the area.[14] By Albian times the transgression was fully established across the entire graben. Upper Aptian to Lower Albian deposition appears to have taken place in a dominantly inner neritic environment with some short-lived continental episodes suggestive of tropical coastal lagoons and deltaic flood plains. Localized tectonism is thought to have occurred during the Albian and is reflected in the development of offshore bar sandstones and, occasionally, limestones. This Greensand interval is well developed in the Kinsale Head Gas Field area.[17] It is approximately 150m thick and typically comprises a basal shore-line sandstone deposit overlain by marine shelf glauconitic claystones. These are followed by a cleaning-upwards sandstone sequence of stacked offshore bars. Predominant sediment transport was from the northeast with bimodal tidal currents reworking, sorting and winnowing the sands. These in turn are succeeded by another grey poorly-indurated bioturbated claystone unit (Fig. 4.6).

The Albian clastics were succeeded in the Cenomanian by the establishment of Chalk deposition throughout the Celtic Sea Grabens. This Chalk succession is up to 1,000m thick and generally spans a Cenomanian to possibly Maastrichtian age. The Chalk crops out extensively on the sea floor over much of the Celtic Sea area (Fig. 4.2). Although the present northern margin of Chalk distribution in the offshore lies just off the south coast of Ireland close to the northern margin of the North Graben, a single occurrence of Upper Campanian Chalk is preserved onshore within a collapsed cave system at Ballydeenlea, near Killarney in County Kerry.[2] The Chalk therefore overstepped the confines of the grabens and a thin layer was probably deposited over most of the southern part of the Irish continental platform area.

Sea bed sampling indicates a general absence, due to erosion following Palaeocene inversion, of Upper Cretaceous Chalk in the eastern part of the North Graben. In the South Graben an extensive Chalk development occurs over the main part of the area (Fig. 4.2).

The Chalk reflects the open marine conditions of the widespread Upper Cretaceous marine transgression. Terrigenous influence was slight to moderate during the early phase of carbonate deposition and the clastic input had ceased entirely by approximately Turonian times. The Chalk itself is largely pure calcium carbonate containing, in addition to some argillaceous material in the lower parts, variable amounts of glauconite, shell fragments and chert.

Tertiary sediments crop out on the sea bed in the Celtic Sea area as a westwards-thickening sequence unconformably overlying Mesozoic strata (Fig. 4.2). They occur on the Pembrokeshire Ridge and also as a tongue-like extension along part of the northern margin of the North Celtic Sea Graben. The Tertiary is poorly represented in the Bristol Channel area, except in the small Stanley Bank Basin in the southwest corner of the East Bristol Channel Basin where borehole evidence indicates the presence of up to 350m of Middle to Upper Oligocene siltstones, claystones and lignites.

A thick sequence of up to 800m of Tertiary claystones, shales, thin and locally dolomitic sandstones and lignites occurs in the St. George's Channel Basin. These range in age from probable Late Palaeocene or Eocene to possibly Neogene. A thin Tertiary sequence overlies St. Tudwal's Arch while 600m of probably Neogene sediments occur in the Mochras borehole at the eastern end of the Cardigan Bay Basin.

The Tertiary in the Celtic Sea Grabens is likely to commence with fluvial, and possibly occasional shallow or marginal marine, deposits overlain by fully marine sediments.

Hydrocarbon Potential

By comparison with most other offshore Irish and West British sedimentary basins, a relatively large number of hydrocarbon exploratory wells have been drilled in the general Celtic Sea area. Although very few of these wells reached economic basement, hydrocarbon shows were encountered in many (Table 4.1), indicating the overall general prospectivity of the area. However, the most promising part of the general Celtic Sea area appears to be the North Celtic Sea Graben, where the Kinsale Head Gas Field, with recoverable gas reserves of the order of 1.35 Tcf, is located. In addition to this, which is the only producing field in the Irish offshore areas, a number of currently non-commercial accumulations have been discovered in the basin but have not been fully appraised. The largest and most promising of these to date is the Seven Heads prospect where two wells flowed hydrocarbons on test. The Esso/Marathon 48/24-1 discovery well flowed at a rate of 780 b/d and produced gas at rates of up to 10 MMcfd. The Esso/Marathon 48/28-1 well produced at a rate of 1,550 b/d on test fron one thin sandstone. A further well, 48/23-1, had oil and gas shows but failed to flow hydrocarbons to surface on test. The final well on the prospect, Esso/Marathon 48/24-2, was drilled in 1978 and extensive Formation Interval Testing showed the presence of both oil and gas zones. Reserve estimates by Esso for the prospect have been put at about 100 Bcf of gas and 1m to 2m barrels of oil.[19]

Gas flow rates of 119 Mcfd were recorded in the Marathon 49/13-1 discovery well on the Ardmore prospect. An offset well, Marathon 49/14-1, flowed gas on test at a rate of 8.81 MMcfd. This prospect, like that at Seven Heads, is currently regarded by the company as being non-commercial. However, these encouraging aspects of the drilling in the area strongly indicate that the North Graben contains appreciable quantities of hydrocarbons, although individual fields may be small by North Sea standards.

A number of both source and reservoir horizons are known to exist within the Celtic Sea area. Thermally mature source rocks containing locally variable poor to very good potential to generate both oil and gas are likely to exist at Lower, Middle and Upper Jurassic levels. The Lower Cretaceous, although expected to be frequently thermally immature, may also contain locally rich source rock potential. If Coal Measures are present in the eastern part of the Celtic Sea and St. George's Channel area they may also represent a potential gas source, although Hercynian deformation may have partly or completely destroyed their source potential. The Lower Cretaceous Wealden and Greensand-equivalent sandstones are the producing horizons in the Kinsale Head Gas Field and have been the prime exploration target in the basin. They still remain attractive as potential reservoir horizons. The Upper Jurassic and probably less common Middle Jurassic sandstones must represent highly prospective horizons while Permo-Triassic continental clastics can also be regarded as potential reservoirs especially close to the margins of the graben and in the Irish Sea area, where they are unlikely to lie at very great depths.

Basin inversion occurred during Upper Cretaceous and Palaeocene times, when the central axis of the North Graben was uplifted to

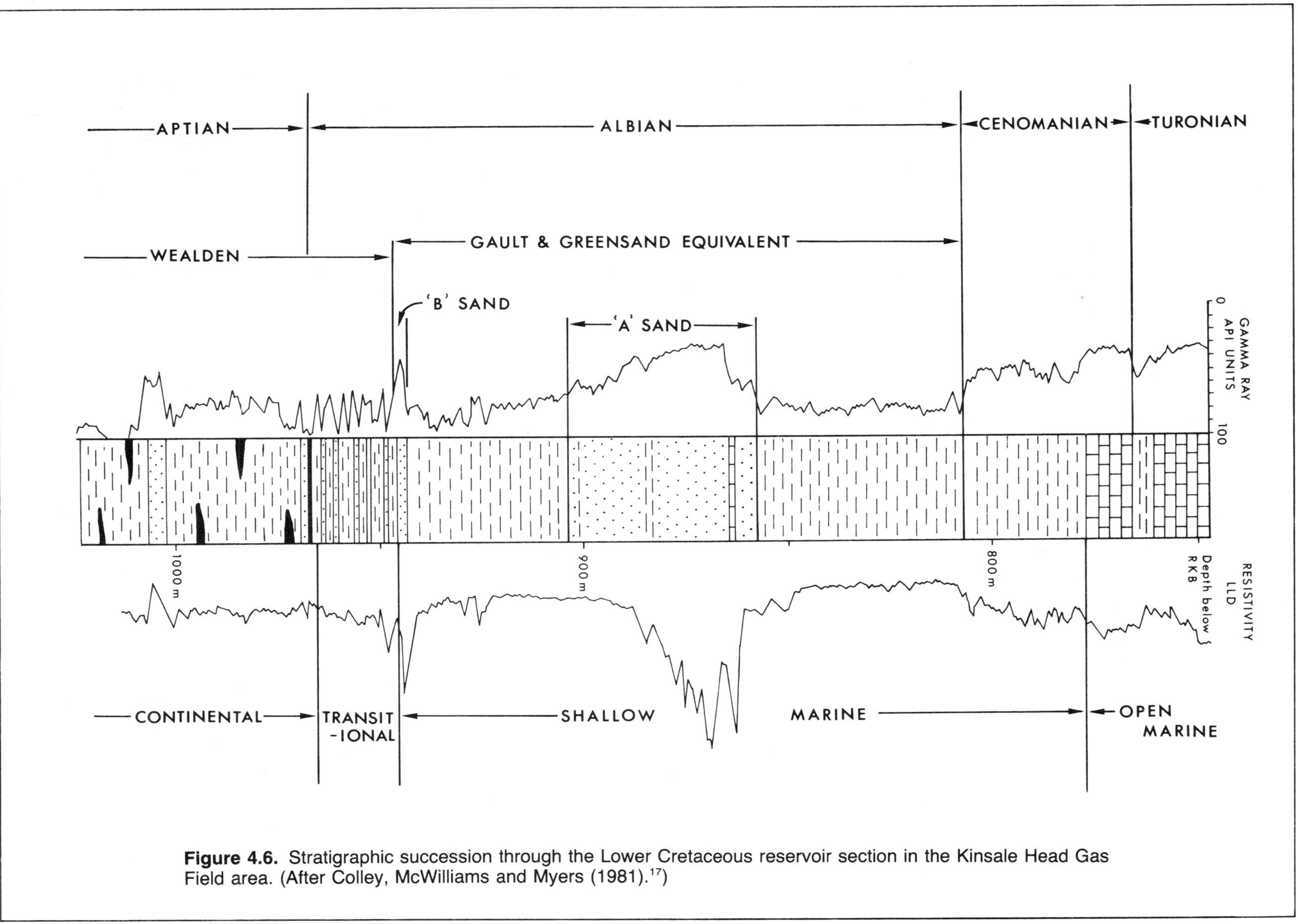

Figure 4.6. Stratigraphic succession through the Lower Cretaceous reservoir section in the Kinsale Head Gas Field area. (After Colley, McWilliams and Myers (1981).[17])

form the Kinsale Head structure. Similar structures could be anticipated along the central axis of the basin. Northeast-southwest and northwest-southeast extensional faults also occur and, although some of these were active until post-Base Upper Cretaceous Chalk times, they could be expected in places to have assisted with the formation of trapping structures. Unconformity traps could also be anticipated in the area, in addition to tilted fault block plays.

TABLE 4.1 Wells drilled in the Irish sector of the Celtic Sea Basins.

Well name	*Area*	*Latitude Longitude*	*Spud date Rig release date*	*R. T. Elevation Water depth*	*Total depth Well status*
Marathon 48/25-1	North Celtic Sea Graben	51°18′55.100″N 08°03′19.900″W	16.05.′70 25.02.′71	9m 94m	3,335m. Gas shows. P & A.
Marathon 50/11-1	North Celtic Sea Graben	51°34′55.170″N 06°57′18.540″W	27.02.′71 10.09.′71	9m 72m	3,629m. Oil and gas shows. P & A.
Marathon 48/25-2	North Celtic Sea Graben	51°19′39.505″N 08°05′48.595″W	13.09.′71 9.11.′71	9m 93m	1,996m. Gas flows (26.45 MMcfd) P. & A.
Marathon 48/20-1	North Celtic Sea Graben	51°21′11.000″N 08°00′11.000″W	19.02.′72 27.04.′72	9m 92m	853m. Not completed for technical reasons. P. & A.
Marathon 48/20-1A	North Celtic Sea Graben	51°21′07.772″N 08°00′21.764″W	6.05.′72 25.07.′72	9m 90m	2,164. Gas flows (13.682 MMcfd) P. & A.
Marathon 58/3-1	South Celtic Sea Graben	50°53′46.693″N 07°35′10.207″W	31.07.′72 23.09.′72	9m 101m	2,859m P. & A.
Esso 56/20-1	North Celtic Sea Graben	50°29′06.624″N 09°11′15.255″W	16.08′72 18.10.′72	9m 129m	2,417m P. & A.
Marathon 49/11-1	North Celtic Sea Graben	51°32′24.963″N 07°51′27.106″W	27.09.′72 22.02.′73	9m 81m	3,647m. Oil shows. P. & A.
Marathon 49/16-1	North Celtic Sea Graben	51°21′38.342″N 07°52′02.513″W	5.06.′73 29.06.′73	9m 90m	1,088m. Gas shows. P. & A.
Marathon 49/16-2	North Celtic Sea Graben	51°23′16.180″N 07°57′34.859″W	4.07.′73 9.08.′73	9m 89m	1,311m. Gas flows (60.178 MMcfd) P. & A.
Esso 48/30-1	North Celtic Sea Graben	51°03′45.000″N 08°05′27.115″N	23.07.′73 11.10.′73	13m 103m	2,910m. Oil and gas shows. P. & A.
Marathon 49/29-1	South Celtic Sea Graben	51°04′24.400″N 07°19′32.800″W	11.08.′73 20.10.′73	9m 97m	3,201m P. & A.
Esso 48/24-1	North Celtic Sea Graben	51°11′27.484″N 08°23′40.487″W	15.10.′73 re-entered 7.05.′74 20.07.′74	13m 103m	2,616m. Oil and gas flows (780 b/d, 10 MMcfd) P. & A.
Esso 56/14-1	North Celtic Sea Graben	50°35′06.613″N 09°15′17.596″W	13.02.′74 25.04.′74	13m 105m	2,827m P. & A.
Marathon 49/13-1	North Celtic Sea Graben	51°35′49.463″N 07°33′35.689″W	13.07.′74 12.10.′74	9m 79m	2,900m. Gas flows (0.119 MMcfd) P. & A.
Esso 48/28-1	North Celtic Sea Graben	51°09′53.080″N 08°27′46.424″W	27.09.′74 26.11.′74	13m 103m	1,752m. Oil and gas flows (1,550 b/d, 0.566 MMcfd) P. & A.
Marathon 49/14-1	North Celtic Sea Graben	51°30′26.605″N 07°21′27.280″W	13.10.′74 re-entered 15.06.′75 1.08.′75	9m 80m	2,427m. Oil and gas shows. P. & A.

TABLE 4.1 – continued

Well name	*Area*	*Latitude Longitude*	*Spud date Rig release date*	*R. T. Elevation Water depth*	*Total depth Well status*
Marathon 50/12-1	North Celtic Sea Graben	51°34′44.037″N 06°42′19.470″W	13.06.′75 7.07.′75	24m 72m	1,317m P. & A.
Esso 56/12-1	North Celtic Sea Graben	50°37′58.642″N 09°46′32.627″W	22.06.′75 26.08.′75	13m 134m	2,735m P. & A.
Marathon 50/11-2	North Celtic Sea Graben	51°34′28.630″N 06°59′22.526″W	09.07.′75 10.08.′75	24m 73m	1,768m P. & A.
Marathon 49/20-1	North Celtic Sea Graben	51°28′47.042″N 07°08′05.755″W	2.08.′75 13.09.′75	9m 79m	2,008m P. & A.
Marathon 49/14-2	North Celtic Sea Graben	51°30′16.351″N 07°16′43.754″W	12.08.′75 20.09.′75	24m 81m	2,030m Oil shows. P. & A.
Esso 47/30-1	North Celtic Sea Graben	51°03′12.251″N 09°02′05.130″W	27.08.′75 11.11.′75	13m 118m	2,616m. Oil and gas shows. P. & A.
Marathon 50/3-1	North Celtic Sea Graben	51°53′48.976″N 06°32′09.875″W	29.04.′76 21.07.′76	9m 71m	2,705m P. & A.
Esso/ Marathon 48/28-1	North Celtic Sea Graben	51°10′48.460″N 08°33′58.359″W	3.06.′76 13.08.′76	13m 104m	2,530m. Oil and gas shows. P. & A.
Esso/ Marathon 57/6-1	North Celtic Sea Graben	50°49′36.863″N 08°49′05.865″W	23.07.′76 8.09.′76	9m 106m	2,487. Oil and gas shows. P. & A.
Marathon 49/30-1	South Celtic Sea Graben	51°00′32.200″N 07°04′00.379″W	8.08.′77 23.08.′77	9m 105m	1,067m P. & A.
Marathon 42/17-1	Carnsore Trough. St. George's Channel Basin	52°28′58.695″N 05°45′48.393″W	9.03.′78 re-spudded 13.03.′78 26.04.′78	25m 75m	1,514m P. & A.
Esso/ Marathon 48/22-1	North Celtic Sea Basin	51°12′35.344″N 08°47′29.731″W	27.04.′78 re-spudded 28.04.′78 8.07.′78	14m 106m	2,788m P. & A.
Esso/ Marathon 48/24-2	North Celtic Sea Graben	51°10′18.620″N 08°14′48.837″W	10.07.′78 31.08.′78	14m 100m	2,244m. Oil and gas shows. P. & A.
Elf 56/9-1	North Celtic Sea Graben	50°42′50.610″N 09°18′56.800″W	28.04.′79 23.06.′79	26m 124m	2,595m P. & A.
Marathon 49/17-1	North Celtic Sea Graben	51°24′18.769″N 07°41′49.880″W	16.05.′79 30.08.′79	25m 85m	2,599m. Oil and gas shows. P. & A.
B.P. 50/12-2	North Celtic Sea Graben	51°39′16.000″N 06°47′14.080″W	1.04.′81 19.04.′81	27m 73m	497m. Not completed for technical reasons. P. & A.
B.P. 50/12-2A	North Celtic Sea Graben	51°39′15.334″N 06°47′13.319″W	19.04.′81 7.08.′81	27m 73m	3,098m. Oil and gas shows. P. & A.

References

1. ZIEGLER, P.A. 1981. Evolution of Sedimentary Basins in North-West Europe: *in* Illing, L. V. and Hobson, G. D. (*Eds*), *Petroleum Geology of the Continental Shelf of North-West Europe.* Heyden & Son Ltd., London, 3–39.
2. WALSH, P. T. 1966. Cretaceous outliers in South West Ireland and their implications for Cretaceous palaeogeography. *Q. J. geol. Soc. London* **122**, 63–84.
3. DAY, G. A. and WILLIAMS, C. A. 1970. Grav-

ity compilation in the N. E. Atlantic and interpretation of gravity in the Celtic Sea. *Earth planet, Sci. Lett.* **8**, 205–213.

4. DAVEY, F. J. 1970. Bouger anomaly map of the North Celtic Sea and entrance to the Bristol Channel. *Geophys. J. R. astron. Soc.* **22**, 277–282.
5. BLUNDELL, D. J. 1979. The Geology and Structure of the Celtic Sea: *in* Banner, F. T. Collins, M. B. and Massie, K. S. (*Eds*). *The North-West European Shelf Seas: the Sea Bed and the Sea in Motion. 1. Geology and Sedimentology.* Elsevier, Amsterdam, 43–60.
6. HOLDER, A. P. and BOTT, M. H. P. 1971. Crustal structure in the vicinity of south-west England. *Geophys. J. R. astron. Soc.* **23**, 465–489.
7. BUNCE, E. T., CRAMPIN, S., HERSEY, J. B. and HILL, M. N. 1964. Seismic refraction observations on the continental boundary west of Britain. *J. Geophys. Res.* **69**, 3853–3863.
8. REEVES, T. J., ROBINSON, K. W. and NAYLOR, D. 1978. Ireland's Offshore Geology. *Irish Offshore Review.* **1**, 25–28.
9. BARR, K. W., COLTER, V. S. and YOUNG, R. 1981. The Geology of the Cardigan Bay – St. George's Channel Basin: *in* Illing, L. V. and Hobson, G. D. (*Eds*). *Petroleum Geology of the Continental Shelf of North-West Europe*, Heyden & Son Ltd., London, 432–443.
10. KAMERLING, P. 1979. The Geology and Hydrocarbon Habitat of the Bristol Channel Basin. *J. Petrol. Geol.* **2**, 75–93.
11. ROBINSON, K. W., SHANNON, P. M. and YOUNG, D. G. G. 1981. The Fastnet Basin: An Integrated Analysis: *in* Illing, L. V. and Hobson, G. D. (*Eds*). *Petroleum Geology of the Continental Shelf of North-West Europe.* Heyden & Son Ltd., London, 444–454.
12. DOBSON, M. R., EVANS, W. E. and WHITTINGTON, R. 1973. The geology of the South Irish Sea. *Rep. Inst. geol. Sci. London*, 73/11.
13. DELANTY, L. J., WHITTINGTON, R. J. and DOBSON. 1981. The Geology of the North Celtic Sea west of 7° Longitude. *Proc. R. Ir. Acad.* **81B**, 37–54.
14. COLIN, J. P., LEHMANN, R. A. and MORGAN, B. E. 1981. Cretaceous and Late Biostratigraphy of the North Celtic Sea Basin, Offshore Southern Ireland: *in* Neale, J. W. and Brasier, M. D. (*Eds*). *Microfossils from recent and fossil shelf seas.* Ellis Horwood Ltd., Chichester, 122–155.
15. GARDINER, P. R. R. and SHERIDAN, D. J. R. 1981. Tectonic framework of the Celtic Sea and adjacent areas with special reference to the location of the Variscan Front. *J. Struct. Geol.* **3**, 317–331.
16. TUCKER, M. E. 1977. The Marginal Triassic deposits of South Wales: continental facies and palaeogeography. *Geol. J.* **12**, 169–188.
17. COLLEY, M. G. McWILLIAMS, A. S. F. and MYERS, R. C. 1981. Geology of the Kinsale Head Gas Field, Celtic Sea, Ireland: *in* Illing, L. V. and Hobson, G. D. (*Eds*) *Petroleum Geology of the Continental Shelf of North-West Europe.* Heyden & Son Ltd., London, 504–510.
18. ALLEN, P. 1981. Pursuit of Wealden models. *J. geol. Soc. London* **138**, 375–405.
19. DAFTER, R. 1981. Irish Oil Prospects. Much depends on Porcupine. *Financial Times.* 12th January.

Chapter 5

Fastnet Basin

Introduction

The *Fastnet Basin* is a small elongate northeast-southwest-trending sedimentary basin measuring approximately 110 km long by 40 km wide. It takes its name from the small Fastnet Rock which lies some 10 km off the southern Irish (Co. Cork) coast. The basin lies to the west of the North Celtic Sea Graben (Fig. 4.1) and the boundary between the two is taken as a northwest-trending line running through blocks 56/16, 56/17 and 56/23, north of which the North Graben widens appreciably. This boundary may coincide approximately with the suggested southeastwards extension of the Gibbs Fracture Zone – Clare Lineament trend described in Chapter 7. Water depths in the basin are slightly greater than in the Celtic Sea Grabens and average some 130m or thereabouts.

The Fastnet Basin is partially fault-bounded, with the boundary faults most noticeable at the northwestern and southwestern edges of the basin. Shallow basement platforms comprised of probable Palaeozoic rocks lie to the north, south and west and are covered by a thin layer of Tertiary and/or Upper Cretaceous strata.

The basin has an extensive history of Mesozoic and Tertiary sedimentation and tectonism, with sediment thicknesses in places exceeding 4.5 km. In broad terms the evolution of the basin occurred in a similar palaeogeographic and tectonic framework to that of the adjacent Celtic Sea Grabens, while certain similarities with the nearby Goban Spur (Chapter 6) might also be expected.

During Late Carboniferous and probably Early Permian times the Fastnet Basin area suffered the effects of compressional tectonics associated with the Armorican orogenic belt centred in France and Germany. This compressional setting was replaced in Late Permian times by a general tensional tectonic regime which heralded the commencement of a series of attempts at the breaking apart of the European and American Continental plates, and which successfully culminated, in Cretaceous times, in the opening of the North Atlantic. Block faulting probably occurred in the Fastnet – Celtic Sea area during the Triassic rifting phase, with accompanying clastic continental sedimentation. However, it is likely that the formation of the Fastnet Basin as a subsiding uniform basinal structure did not commence until approximately mid to end Hettangian (Early Jurassic) time.

The shallow to moderately deep basinal sedimentation was interrupted in Bajocian times by a major uplift or thermal arching event with accompanying igneous intrusions and large scale faulting. This sequence of events, which corresponds in time to the Main Cimmerian Phase[1] seen in most of the adjacent basins, may correspond to the cessation of the first abortive attempt at Atlantic opening in the Fastnet – Celtic Sea area, and to the switching of rifting and attempted sea floor spreading to the western sedimentary basins (Porcupine and Rockall Basins). A further major tectonic event in the Fastnet Basin, in common with most of the other Western European sedimentary basins, is the Late Cimmerian event coincided approximately with the Jurassic-Cretaceous boundary. As in the Celtic Sea Grabens the shallow shelf Chalk deposits of the Upper Cretaceous marine transgression overlie the dominantly continental clastic deposits of the Lower Cretaceous and coincide with the separation of the Irish and eastern North American sedimentary basins. The final major tectonic event in the basin is marked by the Lower to Middle Tertiary Laramide unconformity and is probably synchronous with the Alpine Orogeny in Central Europe.

Geophysical Data

The first exploration geophysics in the Fastnet Basin took place in 1964 when SAPA flew a wide spaced aeromagnetic survey over the entire Celtic Sea area. In 1969, Seiscom Delta shot the first multichannel seismic programme over the Fastnet Basin for Conoco. Since then a very substantial amount of geophysical data, particularly gravity, magnetics and reflection seismics has been acquired. In particular, some thousands of kilometres of deep reflection seismic data have been shot. The large amount of geophysical information, together with the well information, has resulted in the small Fastnet Basin being one of the best understood of the offshore Irish Mesozoic basins. The description of the structure of the basin given in this section relies heavily on the recently published paper on the area by Robinson, Shannon and Young.[2]

Little has been published on the gravity or magnetic patterns of the Fastnet Basin. As most of the main features of the area may be interpreted from the seismic and well data, only the very broad outlines derived from other geophysical data are presented in this chapter. These data demonstrate that the Fastnet

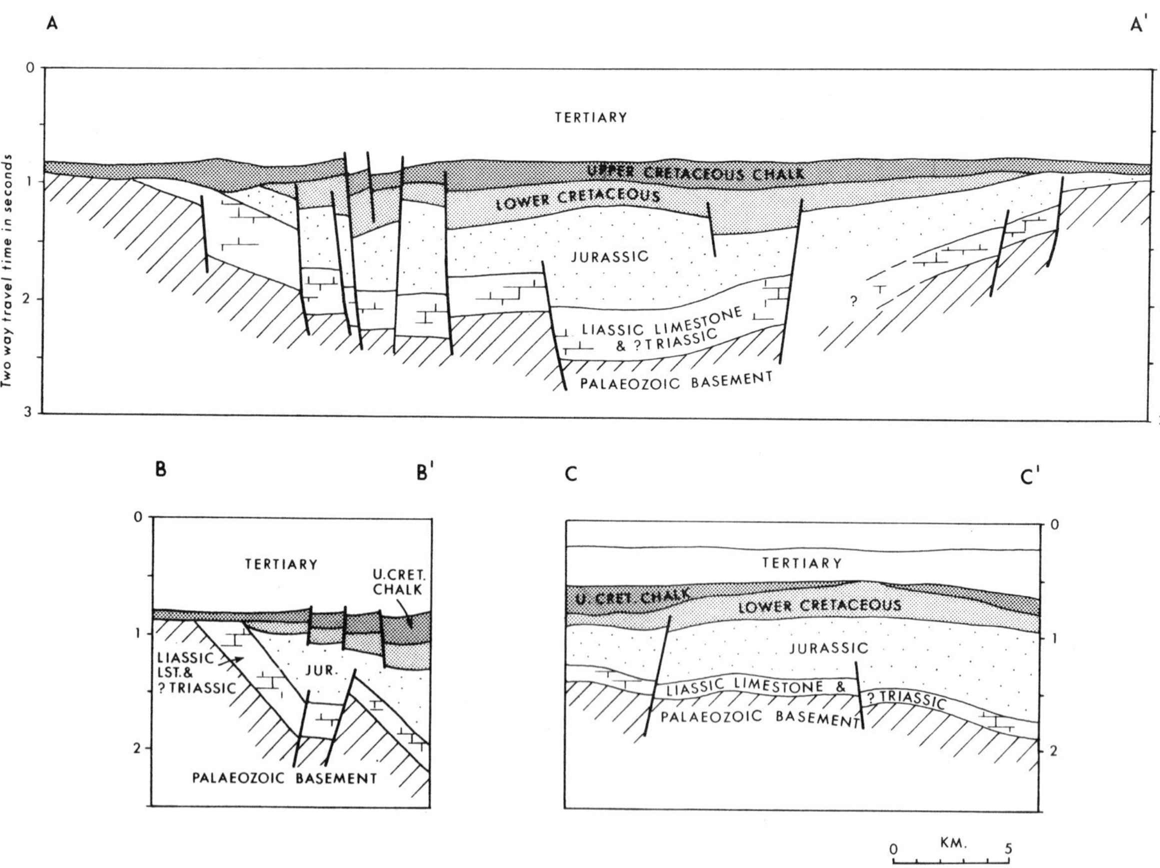

Figure 5.1. Cross-sections, modified from seismic reflection profiles, illustrating various structural features of the Fastnet Basins. (After Robinson, Shannon and Young (1981).[2]) The location of the cross-sections is given in Fig. 5.3.

Basin, in common with all the immediately adjacent basins, is underlain by continental crust. However, some relatively old seismic refraction work[3] suggests that this crust may be stretched and thinned to some 15 km in places under the trough. The gravity and magnetic information fairly clearly reflects the elongate caledonoid trend of the basin and indicates a sedimentary fill of the order of 2–5 km thick. The basin margins are fairly clearly defined. Also highlighted are a number of areas of probable igneous activity especially in the central part of the basin, in the general vicinity of the arbitrary boundary between the Fastnet Basin and the North Celtic Sea Graben and, to a lesser extent, at the southwestern end of the trough.

Much of the tectonic history of the basin is revealed in the seismic profiles of Fig. 5.1, and in the intra-Mesozoic contour maps of Figs 5.2 and 5.3. These two maps, at the Top Liassic Limestone and the base Upper Cretaceous Chalk levels, depict the only two horizons that can be reliably mapped throughout the entire basin. The predominant fault trends on the two maps are similar, although fault displacements in the Cretaceous are significantly less than in the Lower Jurassic. The most noticeable and best developed fault system trends approximately northeast-southwest and shows

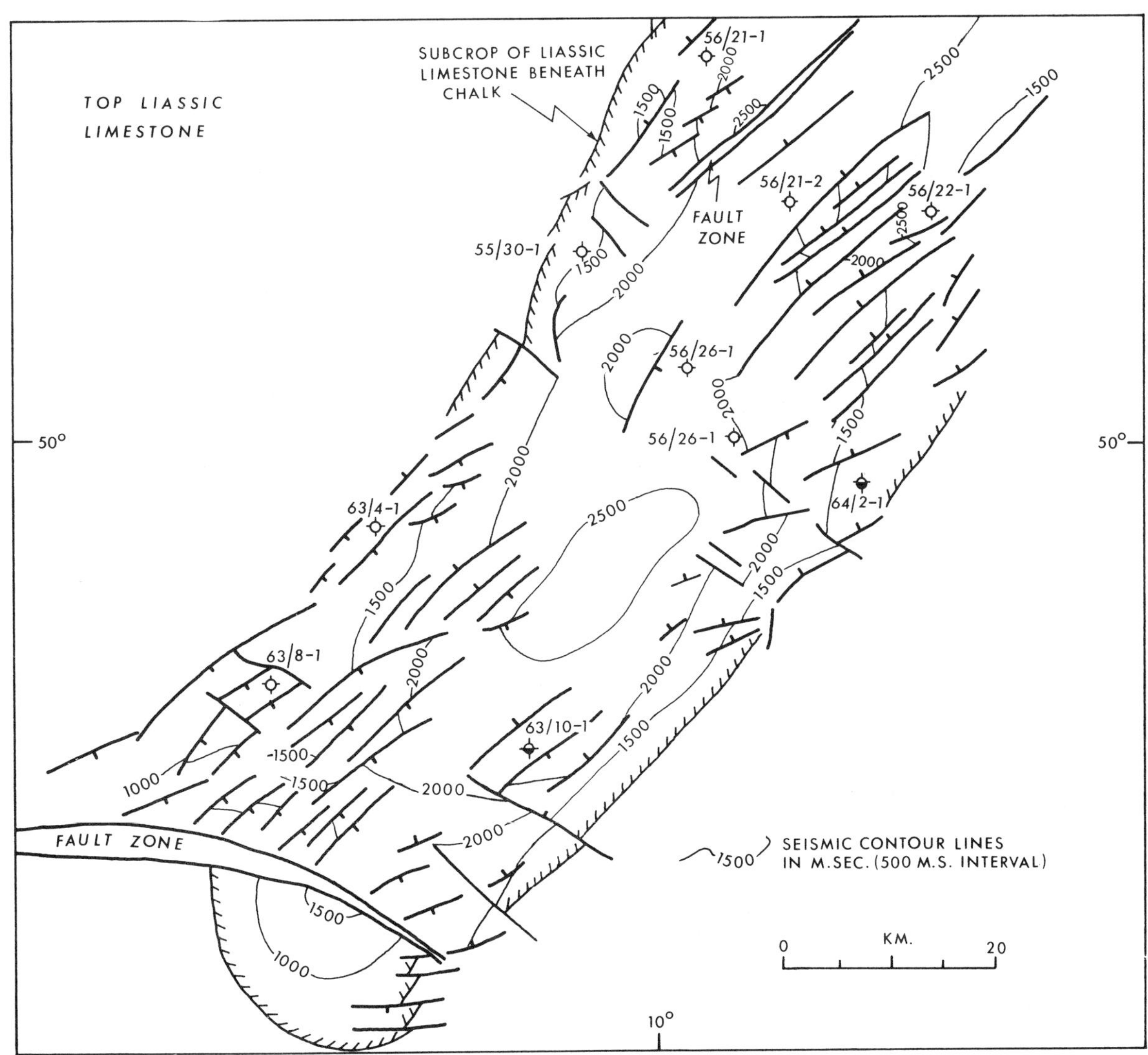

Figure 5.2. Structure contour map of the Fastnet Basin at Top Liassic Limestone level. (After Robinson, Shannon and Young (1981)[2]).

a dip slip, generally down-to-the-basin, sense of movement. This caledonoid direction is probably a reflection of underlying basement trends. Superimposed on this is a secondary series of northwest-southeast faults which are interpreted as frequently having a sinistral strike-slip movement. The secondary fault system is best developed in the centre and southwest of the basin. Faulting at the Top Liassic Limestone level clearly shows the dominant stress system to have been a tensional pull-apart regime, with associated block rotation and wrench faulting.

At the Top Liassic Limestone level, the basin is divided by a northwest-trending fault zone which passes through blocks 55/30 and 64/2-1 (Fig. 5.2). This zone is spatially associated throughout much of its length with the late Bajocian intrusive activity.[2,4] The nature and trend of the basin margins appear to change significantly across this central fault zone (Fig. 5.2).[2] A northwest-southeast fault zone also occurs in the southwest where it effectively forms the southwestern limit of the basin at Liassic Limestone level. Like the central fault zone, this one also appears to have associated Middle Jurassic intrusive activity.

The Base Upper Cretaceous (Chalk) horizon (Fig. 5.3) in general appears to overstep the basin margins as defined by the termination of the Liassic Limestone horizon of Fig. 5.2. The major basement-controlled fault zone at the southwest of the basin is still present at this higher level. The only other boundary fault with displacement at this level occurs in the southern part of the northwest margin. A large number of the faults mapped at the Lower Liassic level can be traced into the Upper Cretaceous, but the effect at the higher level is usually slight. Some of the faults at the higher level do, however, display large displacements and the majority of these show evidence of throw reversal as traced from the lower to the higher level (Fig. 5.1). At this horizon there is a widespread positive area extending through blocks 63/4 and 63/8 on the southeastern side of the main boundary fault. The Chalk is entirely absent over the crest of this positive and anomalous area in block 63/8, and its absence re-emphasizes the uplift seen at Liassic Limestone level (Fig. 5.3). Elsewhere in the basin gentle folding and buckling can be seen on the Base Chalk horizon.

A number of instances of basin inversion are seen on seismic sections throughout the Fastnet Basin.[2] In many cases this phenomenon is seen to be fault-related, while in others the inversion appears to be independent of faulting. In general, the main phase of basin inversion occurred in Lower Cretaceous time, and such effects are not usually recorded at Upper Cretaceous level to any marked degree.

The overall orientation of the basins in the Fastnet-Celtic Sea area, with a northeast-southwest trend in the Fastnet Basin and the St. George's Channel Basin and an east-northeast trend in the main North Celtic Sea Graben suggests that the tectonic evolution of the area took place in a dextral rotational simple shear regime. This was possibly induced by the stresses which led to the opening of the Bay of Biscay in Cretaceous times. The northwest-southeast sinistral strike-slip faults in the Fastnet Basin are interpreted,[2] as representing the antithetic shear direction of the strain ellipse. The predominant normal northeast-southwest faults are developed in the tensional sector of the strain ellipse, along a pre-existing caledonoid trend while the sub-parallel synthetic shear direction is largely suppressed. The observed slight strike swing of many of the faults in the area may be due to the clockwise rotation of the strain ellipse as deformation proceeded.

The main sinistral strike slip movement of the northwest-southeast faults, together with the rhombic shapes of the fault-bounded blocks in the basin, is likely to have induced a counterclockwise rotation of the fault blocks, with localized tension developing along the faults. This tension may have facilitated the intrusion of the Middle Jurassic dolerites and micro-gabbros along some of the major faults of this trend.

Stratigraphy

The first exploration drilling in the Fastnet Basin commenced in April 1976. Between then and September 1978 a total of nine wells were drilled throughout the basin (Table 5.1). Two wells each were drilled by Elf Aquitaine, BP and Deminex while Ranger, Cities Service and Texaco drilled one well each. A tenth exploration well (Cities Service 63/10-1) was drilled in 1981 and, although the geological results of that well are confidential, scouting reports suggest that it bottomed in the Triassic. A moderately good geographic spread of wells throughout the basin therefore exists and the generalized stratigraphic compilation for the first nine wells is given in Fig. 5.4. Most of the stratigraphic information given in this section

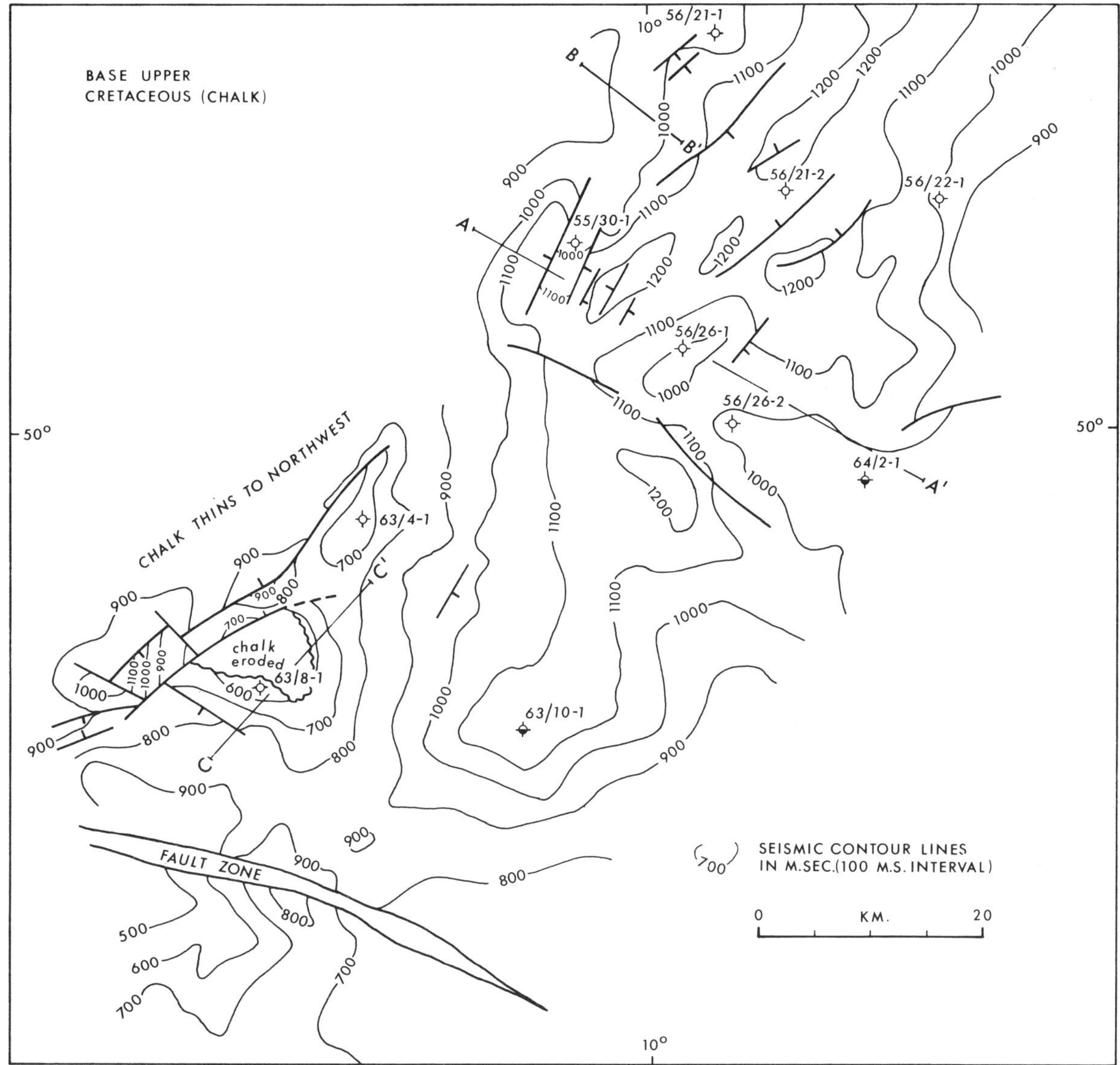

Figure 5.3. Structure contour map of the Fastnet Basin at Base Upper Cretaceous Chalk level. (After Robinson, Shannon and Young (1981).[2])

is based on the paper by Robinson, Shannon and Young.[2]

Upper Palaeozoic rocks form the economic basement of the Fastnet Basin and have been penetrated in two of the wells. Elf Aquitaine 55/30-1, the first well drilled in the basin, bottomed in Devonian (possibly Frasnian) continental redbeds and tuffs, while Cities Service 63/4-1 bottomed in Lower Carboniferous (Late Tournaisian) shelf limestones. The latter resemble those which crop out onshore in Ireland north of an east-west line through Cork city (Fig. 3.2). South of this line non-carbonate basinal fine-grained clastics (Cork Beds) occur, similar to those of at least part of the North Celtic Sea Graben (see Chapter 4). This may suggest that either a complex tongue of shelf sedimentation extended between onshore Ireland and the general vicinity of the 63/4-1 well, or more likely that the area of the well lay on the southern (or western) carbonate shelf of a Carboniferous sedimentary basin of which onshore Ireland formed the northern flank.

In wells 55/30-1 and 63/4-1 economic basement is unconformably overlain by rocks which have a characteristic Triassic lithology.

TABLE 5.1 Wells drilled in the Fastnet Basin

Well Name	*Latitude Longitude*	*Spud date Rig release date*	*R. T. elevation Water depth*	*Total depth Well status*
Elf 55/30-1	50°09′09.630″N 10°05′54.190″W	18.04.'76 28.06.'76	24m 130m	2,800m P. & A.
B.P. 56/26-1	50°03′56.180″N 09°57′54.329″W	1.07.'76 17.08.'76	24m 141m	2,931m P. & A.
Deminex 56/21-1	50°18′54.660″N 09°55′14.060″W	16.07.'76 26.08.'76	25m 130m	2,850m P. & A.
Ranger 63/8-1	49°48′16.226″N 10°29′54.908″W	25.04.'77 21.05.'77	27m 136m	1,920m P. & A.
Cities Service 63/4-1	49°55′56.567″N 10°21′52.123″W	4.06.'77 7.07.'77	33m 128m	1,645m P. & A.
Deminex 56/21-2	50°11′30.809″N 09°49′03.655″W	4.04.'78 23.06.'78	27m 125m	3,720m P. & A.
B.P. 56/26-2	50°00′25.664″N 09°53′52.248″W	22.04.'78 7.07.'78	29m 145m	3,136m P. & A.
Elf 64/2-1	49°57′40.884″N 09°43′47.004″W	2.07.'78 5.08.'78	26m 133m	2,252m Oil shows P. & A.
Texaco 56/22-1	50°11′08.551″N 09°38′10.082″W	7.08.'78 5.09.'78	26m 115m	2,275m P. & A.
Cities Service 63/10-1	49°45′45.099″N 10°10′16.098″W	21.05.'81 5.08.'81	33m 143m	3,447m Oil shows P. & A.

It is worthy of note that nowhere in the basin have proven Permian sediments been encountered. Dominantly continental, probably Triassic, sediments appear to be developed throughout the basin, with a maximum thickness of the order of 600–700m indicated from well and seismic information. Keuper-type sediments are thought to be preserved throughout the basin, with the possible exception of the anomalous fault-bounded area, referred to above in the Geophysical Data section, in block 63/4. However, the preservation of Bunter-type sediments appears to be more localized and they have probably been encountered in only two of the wells (Cities Service 63/4-1 and BP 56/26-2) drilled to date. The Bunter, where present, is represented by continental sandstones and lacustrine and sub-tidal to supra-tidal marls. It probably represents sedimentation in localized small fault-controlled basins similar to that occurring in the North Celtic Sea Graben (see Chapter 4). In general the Keuper lithologies comprise an arenaceous sequence overlain by argillaceous sediments. The sandstones generally reflect coarse aeolian and flash-flood fluviatile deposition, but some coastal marine influences are indicated by the presence of occasional interbedded carbonates. This sequence generally passes up into evaporitic shaly marl deposits indicative of deposition in a sabkha or marginal lacustrine depositional environment.

Triassic continental sedimentation gave way in Rhaetian times to carbonate formation, characteristic of the widespread Lower Liassic marine transgression. In the centre of the basin this carbonate sedimentation, which gave rise to the well-defined Liassic Limestone formation, continued until end-Hettangian times while in the south of the basin it lasted until the early Sinemurian. Over northern parts of the basin stable shelf conditions persisted until the end of lower Sinemurian times. In general, the Liassic Limestone averages 250m in thickness but may be as thick as 460m. The presence of brackish and freshwater microfaunas, with periodic penetration by coastal marine facies towards the base of the carbonate succession, indicates an unstable

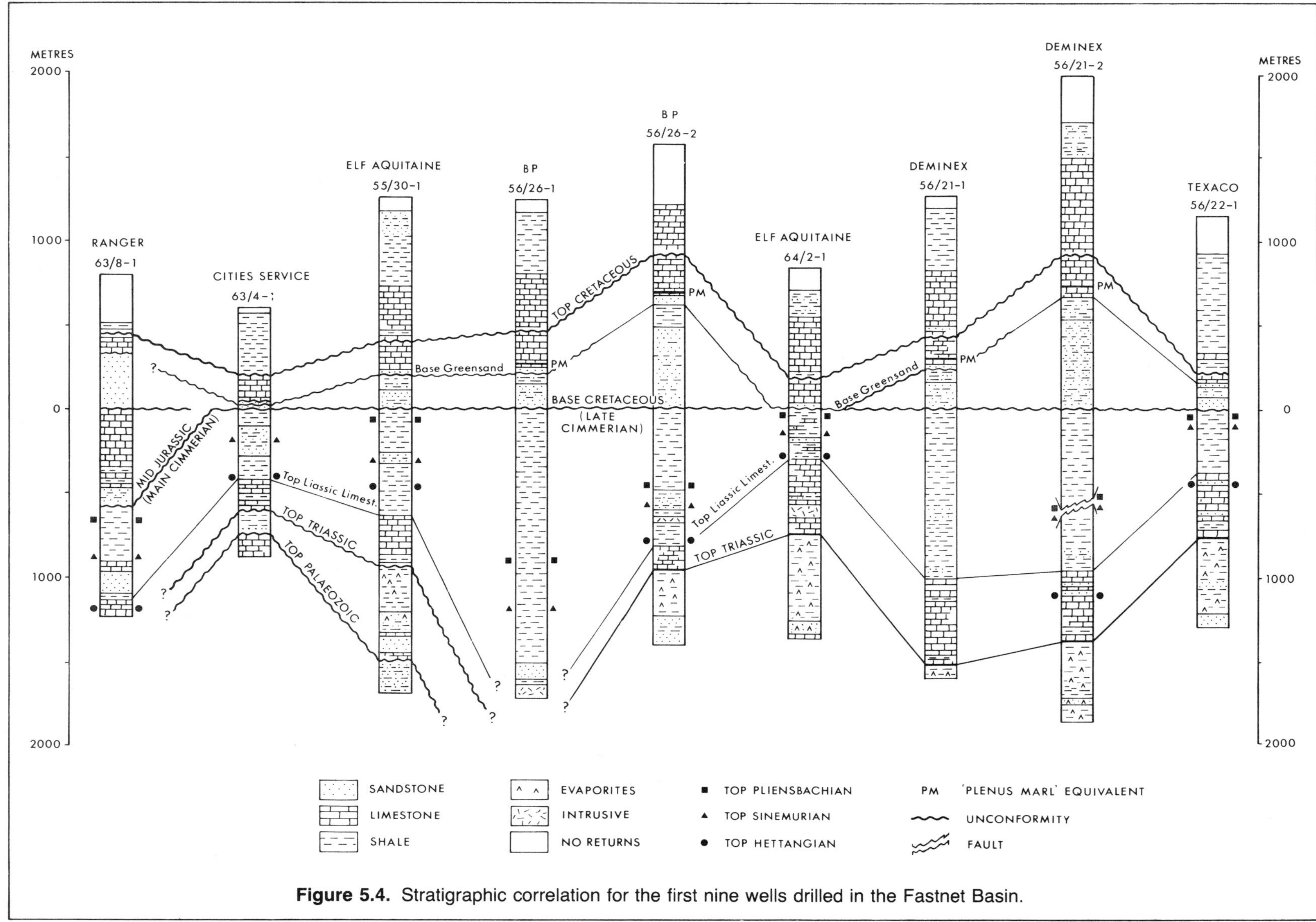

Figure 5.4. Stratigraphic correlation for the first nine wells drilled in the Fastnet Basin.

marginal environment at the commencement of the marine transgression. The brackish aspect of the limestones was rapidly lost upwards through time, and by Hettangian time the Jurassic transgression was well established through the basin. The limestones themselves are sometimes oolitic, suggestive of very shallow marine, turbid, carbonate-rich conditions, and are often interbedded with thin claystone, siltstone or sandstone stringers.

The shelf limestones give way to a thick and usually marly sequence of Sinemurian, Pliensbachian, and often Toarcian and Aalenian, inner to outer shelf claystones. A major sandstone member, up to 230m thick, of Sinemurian to Pliensbachian age occurs in the southern half of the basin. These delta fringe sandstones, with clastic input from the general southwestern quadrant, mark a regressive episode in the Jurassic marine transgression and are diachronous in nature. The sandstones wedge out northeastwards and are not present in the three most northerly wells in the basin. This regressive episode is succeeded by a return to the tranquil outer shelf marly claystones.

Middle or Upper Jurassic sediments were encountered in only two wells. In the BP 56/26-1 well Bajocian mudstones conformably overlie the fine-grained Liassic sediments. In the Ranger 63/8-1 well an unconformity above the Toarcian is followed by a thick probable Bathonian sandstone sequence and a brackish to freshwater thick carbonate succession of Kimmeridgian to Purbeckian age.

The onset of the Cretaceous is everywhere marked by a major unconformity reflecting the Late Cimmerian Phase.[1] A major threefold division of the Cretaceous is obvious within the basin.

The lowermost Cretaceous sequence, which is probably Wealden-equivalent, represents a return to continental deposition and is everywhere typically developed with a lower sandstone sequence followed by an upper claystone sequence. This is broadly similar to the Lower Cretaceous sedimentation pattern observed in the western part of the North Celtic Sea Graben. Although palaeontological control is extremely poor, especially in the lower part of the sequence, the Wealden-equivalent in the Fastnet Basin appears in general to span a Berriasian to possibly Aptian time range.

The lower sandstone sequence ranges in thickness from less than 100m to more than 500m and is everywhere continental in aspect. Typical fluvial and alluvial depositional regimes are present, as illustrated by stacked channel sands, point bar sequences and overbank features. The main source area for the Wealden sands lies on or close to the margins of the basin and in general tended to be in the western quadrant of the area. Occasional brackish to littoral marine microfauna make an appearance towards the top of the Wealden sandstone succession, indicating the onset of Cretaceous subsidence and the associated second major Mesozoic marine transgression. A gradually deepening depositional environment is reflected in the succeeding Wealden claystones which pass upwards from marsh or lagoonal sediments with barrier bar sequences and thin lignites developed, into restricted littoral and neritic marine environments. The source areas continued to lie to the northwest, west or southwest, that is in the general western quadrant.

The onset of the second main Cretaceous sequence, comprising Greensand-equivalent sandstone and minor claystone deposition, is marked on the margins of the basin by the development of a disconformity or minor unconformity. In the basin centre the passage upwards from the Wealden claystone succession is conformable. The Greensand-equivalent interval is frequently diachronous, and ages range from Barremian and Aptian to Cenomanian. It represents a regressive phase in the general Cretaceous transgression, and facies range from coal-bearing deltaic and transitional, to marine inner sublittoral environments. A sediment source area probably lay to the south-southeast of the basin during Greensand deposition.

The third major Cretaceous sequence in the Fastnet Basin is represented by a generally constant thickness (120–160m) of Chalk. The Upper Cretaceous Chalk marine transgression extended beyond the limits of the Early and pre-Cretaceous margins of the Fastnet Basin (Fig. 5.1). The Chalk sequence spans a Cenomanian to Campanian time range with Maastrichtian locally preserved. In the centre of the basin it may be subdivided into lower and upper units separated by a characteristic thin marly band which may represent a Plenus Marl equivalent. The Lower Chalk tends to be slightly arenaceous and was deposited in a nearshore inner shelf environment whereas the Upper Chalk was deposited in more outer shelf conditions.

The Upper Cretaceous Chalk succession is unconformably overlain by Tertiary sediments. These generally commence with a limestone which is of the order of 350m thick throughout most of the basin but is thin or absent over some areas in the southwest. The limestone is of mid-Eocene through Oligocene age. It is succeeded by Miocene and Pliocene claystones which contain some clastic intercalations towards the top. The Tertiary depositional environment was marine inner to outer sublittoral, deepening towards the top of the Oligocene and giving way to open marine shelf outer neritic, and even bathyal, environments in Late Oligocene and Miocene times.

Basic intrusive rocks occur in the central part of the basin. They have been encountered in three wells and reach thicknesses in excess of 40m. Interpretation of seismic records[4] suggests that they have a widespread occurrence throughout the central region of the basin, and appear to be associated mainly with a major northwest-southeast fault zone.[2] The sills have thin chilled margins and have altered and metasomatized the surrounding sediments to a distance of up to 500m. Petrographically they are olivine micro-gabbro and olivine dolerites. Potassium/Argon dating on one sill indicates a probably Late Bajocian age of crystallization which is approximately synchronous with the main period of deformation and the development of the ? mid-Jurassic (Main Cimmerian) unconformity seen in the 63/8-1 well.

Hydrocarbon Potential

Although the first nine wells were plugged and abandoned as dry holes, some recorded minor oil shows, the most significant being in the Liassic sandstones of the Elf 64/2-1 well. The tenth well, Cities Service 63/10-1, encountered oil shows and scouting reports suggest that small quantities of oil were recovered during Drill Stem Testing of a Liassic sandstone interval.

The results from the wells drilled to date indicate that four stratigraphic intervals within the basin contain good quality reservoir sandstones:

1. The *Wealden/Greensand*, where net thicknesses are from 50m to over 140m. Porosities are generally in the range 25–30%, but are occasionally as high as 40%.
2. *Middle Jurassic* sandstones which probably occur in at least part of the southwest of the basin. Over 20m of net sand occur with porosities in the range of 10–15%.
3. The *Sinemurian-Pliensbachian* sandstones in the southwestern half of the basin contain 11–70m of net sand with a general porosity range of the order of 19–38%.
4. *Triassic* sandstones occur with the thickness of good sandstones ranging from 26m to more than 50m. Porosities are generally up to 13%. Within the Fastnet Basin, carbonates are generally non-prospective.

The Liassic Limestone is everywhere tight, with porosities rarely, if ever, exceeding 5%. The Upper Cretaceous Chalk has very good porosities in places, but generally has thin cover, is not close to good nature source rocks and lacks closure. No live oil shows have been recorded in the Chalk.

The available well results indicate that the Liassic shales in the basin contain intervals with locally good to very good potential for oil and gas. However, only in limited areas of the basin are they thermally mature and adjacent to good reservoirs. Nevertheless it is likely that the Liassic shales represent the only significant potential source rock sequence within the basin.

The best potential for hydrocarbon accumulations is likely to exist in the less explored deeper southeastern part of the basin. It is here that the Liassic shales are most likely to have generally reached thermal maturity, good reservoir sandstone sections could be developed in the Lower Cretaceous, Liassic and possibly locally in the Middle Jurassic, while closure may occur in association with the major down-to-the-basin faults which exist in that part of the basin.

References

1. ZIEGLER, P. A. 1981. Evolution of Sedimentary Basins in North-West Europe: *in* Illing, L. V. and Hobson, G. D. (*Eds*). *The Geology of the Continental Shelf of North-West Europe*. Heyden & Son Ltd. London, 3–39.
2. ROBINSON, K. W., SHANNON, P. M. and YOUNG, D. G. G. 1981. The Fastnet Basin: An Integrated Analysis: *in* Illing, L. V. and Hobson, G. D. (*Eds*). *The Geology of the Continental Shelf of North-West Europe*, Heyden & Son Ltd. London, 444–454.
3. BUNCE, E. T., CRAMPIN, S., HERSEY, J. B. and HILL, M. N. 1964. Seismic refraction observations on the continental boundary west of Bri-

tain. *J. geophys. Res.* **69**, 3853–3863.
4. CASTON, V. N. D., DEARNLEY, R., HARRISON, R. K., RUNDLE, C. C. and STYLES, M. T. 1981. Olivine-dolerite intrusions in the Fastnet Basin. *J. geol. Soc. London* **138**, 31–46.

Chapter 6

Goban Spur

Introduction

The *Goban Spur* is a plateau area which lies on the continental margin zone 250 km southwest of the Irish mainland. It is located to the south of the Porcupine Basin, to the southwest of the Fastnet Basin and close to the southwestward extension of the Cornubian Massif which separates the South Celtic Sea Graben from the Western Approaches Basin (Fig. 6.1). The continental - oceanic crustal boundary, although rather poorly delineated at present, appears to lie immediately to the west of the area and trends in a northwesterly direction towards the southern part of the Rockall Trough (Fig. 6.2).

Bathymetrically the Goban Spur comprises an east-trending smooth platform which dips gently westwards from the Celtic Shelf and is cut by very few submarine canyons. Water depths on the platform range from 200m to 2,000m. To the north the platform drops off into the deep part of the Porupine Basin. To the west the gradual slope terminates abruptly at approximately the 2,000m bathymetric contour beyond which the water depth increases dramatically into the Porcupine Abyssal Plain in a southwesterly direction normal to the major northwest-trending fault scarps which govern the tectonic 'grain' of the area (Fig. 6.2).

Although little is known about the geology of the Goban Spur, it is likely, in common with the adjacent basins, to contain a sedimentary sequence of Mesozoic and Tertiary age. However, unlike many of these other areas it is probable that in general the Mesozoic sedimentary cover on the Spur is relatively thin and that the area is structurally rather complex. Extensive faulting appears to dominate the area with the result that there are probably a number of interesting block-fault related small, but possibly thick, sedimentary pockets. The Goban Spur area is unique in its location with respect to the other Irish basins, in that it represents a possible proto-triple point junction area, with the Porcupine Basin extending away northwards, the Fastnet and Celtic Sea Basins trending approximately northeastwards and the Armorican Basin extending southeastwards.

During much of its development history, and in particular up to the end of Cretaceous time, the Goban Spur area was adjacent to the Flemish Cap and Orphan Knoll areas of eastern Canada (Fig. 1.3). As a result, a number of similarities in sedimentation history and evolution between the two areas could be expected.

The eastern part of the Goban Spur, unlike most of the adjacent basins to the north and east, but in common with Flemish Cap which is its mirror in an Irish/Canadian pre-spreading reconstruction, appears to have an underlying Hercynian granite core. As a result the area remained for most of its history in a relatively elevated position, receiving a consequentially thin sedimentary cover. This elevated area appears to have shielded the subsiding western part of the Goban Spur from the normal downslope allochthonous turbidite, slide,

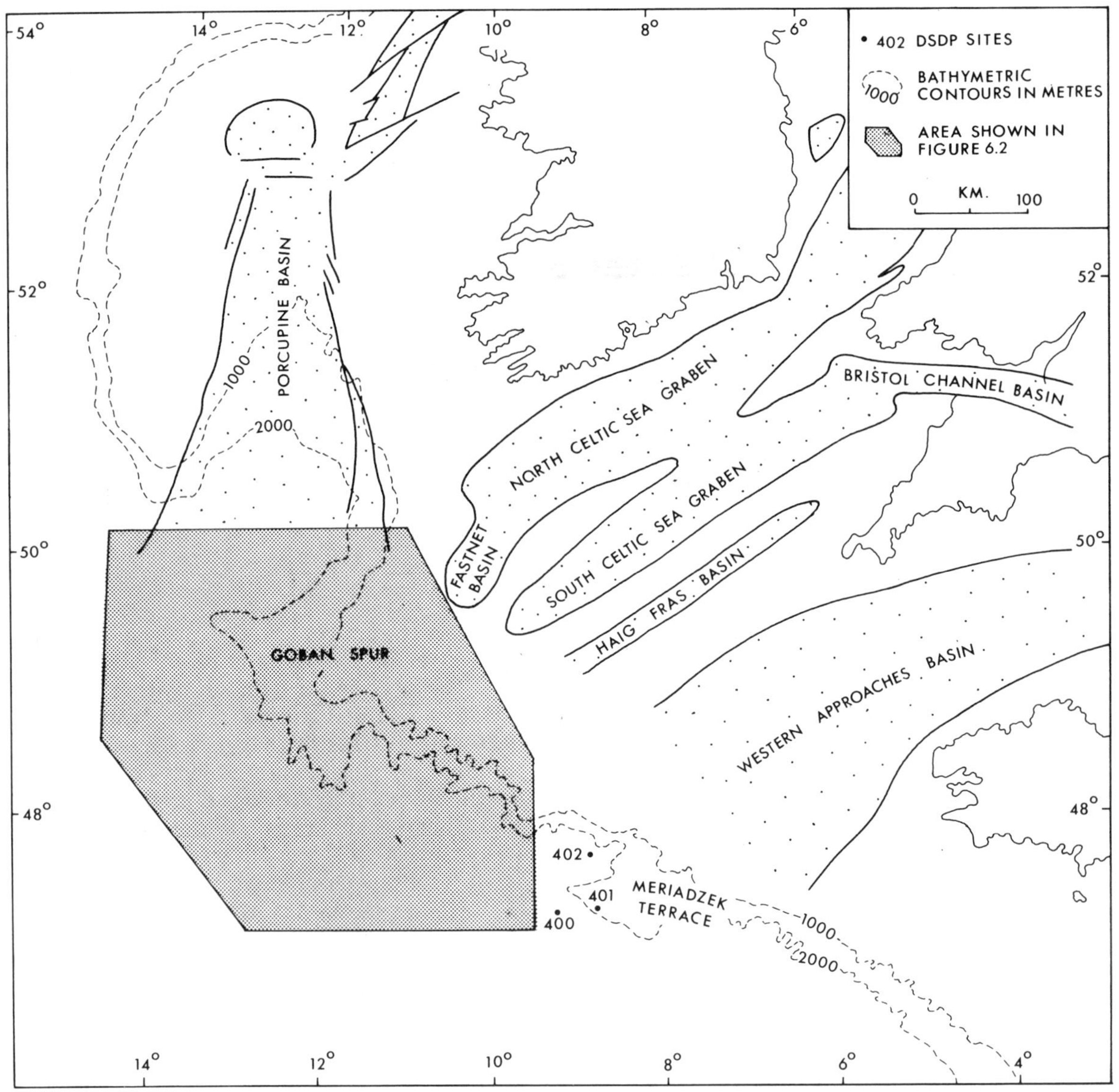

Figure 6.1. Sketch map showing the location of the Goban Spur in relation to adjacent sedimentary basins.

slump and related sediments which would be expected in such a marginal plateau area.

The sedimentation history of the area probably commenced in Permo-Triassic times when the tensional rifting stresses following the culmination of the Hercynian orogeny produced the fault-bounded basins off the coasts of mainland Ireland. The initial syn-rift continental sediments probably gave way to marine deposition in Liassic times as the widespread Jurassic marine transgression became well established. Mid- and especially Late-Jurassic tectonic events, synchronous with the commencement of the opening of the Bay of Biscay, are likely to be represented by at least one Cimmerian unconformity succeeded by Lower Cretaceous continental to marine sedimentation. These in turn should be overlain by fully marine sediments of Upper Cretaceous and Tertiary age representative of post-rifting deposition.

The Goban Spur represents the foremost frontier territory of the Irish offshore hydrocarbon exploration programme. In 1980 a total of 40 blocks covering most of the Spur, and the first to be awarded in the area, were granted to Esso and Amoco/Mobil on an option basis whereby extensive seismic surveys were carried out by the licensees. Esso have now licensed 14 of their 30 option blocks and will drill a well in the area in 1982. However, Amoco/Mobil have decided not to exercise their option and have relinquished their 10 blocks in the eastern part of the Goban Spur.

Geophysical Data

Only reconnaissance geophysical information is currently available on the Goban Spur. Most of this information comes from recently published results of a number of surveys carried out by academic institutions, most notably Edinburgh University and the Institute of Oceanographic Sciences.[1,2] These surveys gathered much valuable information on gravity, magnetic and seismic patterns in the Goban Spur and adjacent areas. In addition approximately 8,000 km of multi-channel seismic have been shot by Esso and Amoco/Mobil during the past two years but these data remain confidential.

Crustal models constructed for the Goban Spur area, based chiefly on the extrapolation of very limited seismic refraction information, suggest that the continental crust which underlies the general area has been only slightly stretched and thinned and is in the thickness range of 16–28 km.[3] Crustal thicknesses of about 22 km have been indicated underlying the main part of the Spur.[4] This figure is similar to the general thickness postulated for the Fastnet Basin to the east (Chapter 5), but is substantially greater than that indicated for the Porcupine Basin to the north where extensive stretching and thinning of the continental crust has been indicated (Chapter 7).

The free-air gravity anomaly pattern suggests a thin sedimentary succession overlying shallow basement. A number of linear anomalies are recognized which probably correspond to underlying northwest-southeast tilted blocks. West of the Spur, a northwest-trending strong positive anomaly, lying oceanwards of a negative anomaly, marks the western edge of the area and corresponds fairly closely with the continental–oceanic crustal transition.

The magnetic anomaly pattern likewise highlights the main northwest-southeast fault patterns in addition to the east-northeast–west-southwest fault system. West of the Goban Spur, strikingly linear northwest-southeast magnetic anomalies characterize the oceanic crust and again fairly clearly define the continental-oceanic crustal boundary.

The geophysical information, and especially the most recent two channel and multi-channel deep seismic surveys, reveal a complex and largely fault-controlled picture of localized sedimentary wedges up to 4 km thick, overlying tilted basement blocks (Fig. 6.3). The basement blocks were originally believed to be massive in nature and seismically opaque.[1] However, more recent seismic information has revealed that large portions of the basement are layered and that such layered sequences are locally in excess of 2 sec in thickness and are occasionally seen to rest unconformably on opaque, possibly true, basement.[2] Two principal and almost orthogonal fault directions occur, one trending east-northeast – west-southwest and the other, and probably older system, running approximately northwest-southeast.

The area is separated from the deeper southern part of the Porcupine Basin by one or more faults, which lie approximately along strike from the northern boundary fault system of the South Celtic Sea Graben. The southern boundary of the Goban Spur is probably formed by a similarly-trending fault or fault zone which is a strike extension of the main northern boundary fault of the Western Approaches Basin. A third major east-northeast – west-southwest fault, crosses just north of the centre of the Spur.[1]

The northwest-striking faults are more prolifically developed than the east northeasterly set, although when traced southeastwards towards the Armorican Basin the latter direction becomes the predominant trend (Fig. 6.2). Faults with these trends form the western and eastern boundaries of the Goban Spur. They are invariably normal and listric faults with southwesterly throws. The amount of displacement decreases from the west of the Spur, where they average up to 3 km, towards the east where the throws are far less impressive. The faults are generally confined to the basement and rarely penetrate to the Upper Mesozoic.

Stratigraphy

No wells have been drilled in the Goban Spur area to date and therefore the age and lithofacies of the sedimentary succession are unknown. However a picture of the likely sequence may be gained by correlation of the seismic stratigraphy with the succession encountered particularly in wells within the nearby Fastnet Basin[5] and to a lesser extent by that described from the Meriadzek Terrace

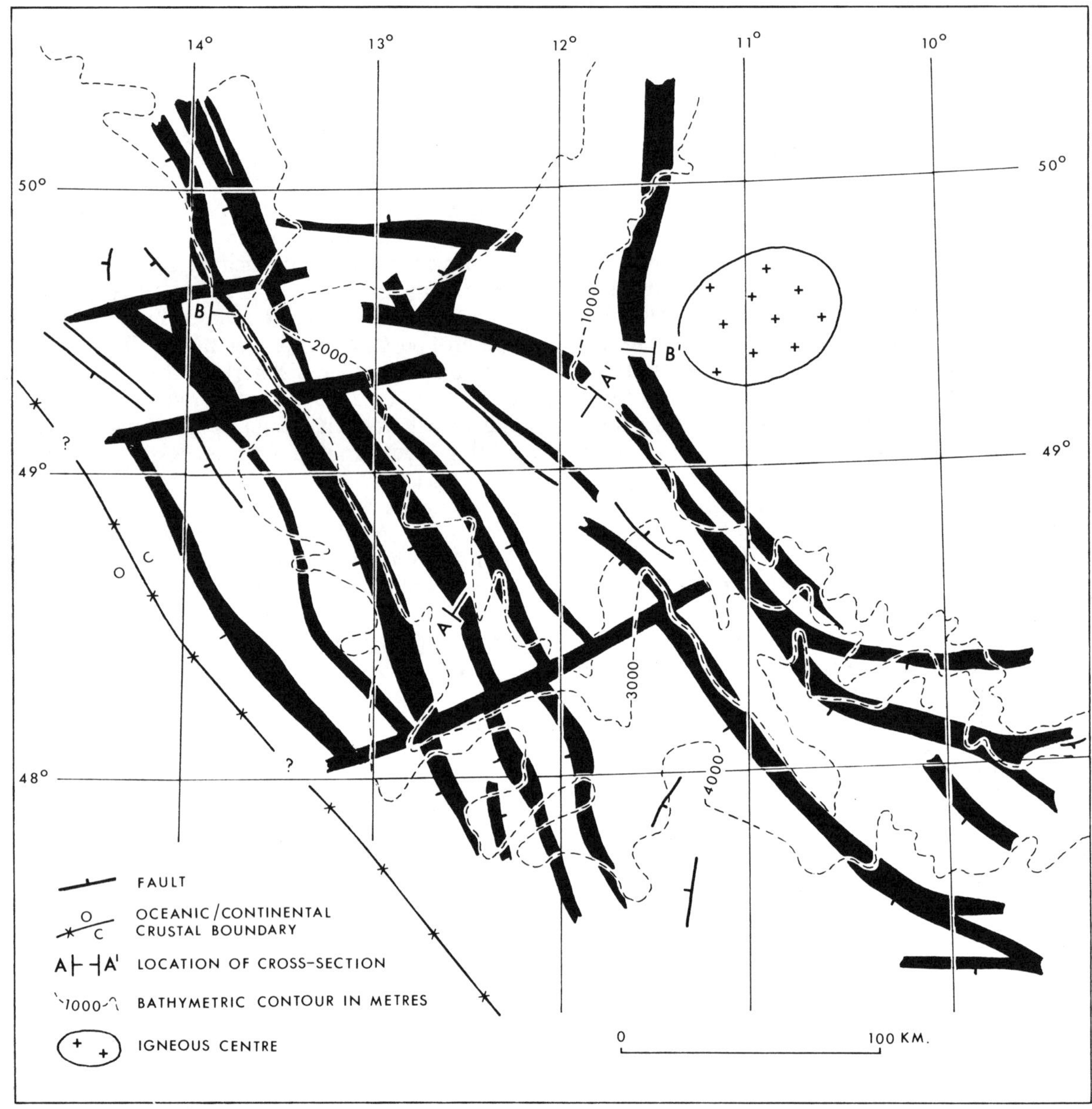

Figure 6.2. Structural map of the Goban Spur showing location and orientation of main faults. (Modified from Dingle and Scrutton (1979)[1] and Roberts *et al.* (1981).[2])

adjacent to the Bay of Biscay at Sites 400A, 401 and 402 on Leg 48 of the Deep Sea Drilling Project[6,7] (Fig. 6.1). In addition to this well information some limited bottom sampling from the Goban Spur[1] has helped to cast further light on the likely succession in the area. However, in making correlations and extrapolations from adjacent areas it must be remembered that the distances involved are considerable (270 km in the case of the DSDP sites) while the tectonic setting of the Goban Spur is different from the two other areas. Nevertheless some similarities should exist in the successions, especially in the post-rifting strata.

The succession in the Goban Spur, as mentioned in the Geophysical Data section above, comprises a tilted fault-block 'basement' which is overlain by four broadly correlatable sedimentary sequences. These are separated by prominent reflecting horizons which represent regional unconformities or disconformities.

Although most workers agree that the post-fault block strata may be subdivided into four sequences, no agreement exists on the ages of all of these units.[1,2]

The fault-block 'basement' is composed of an upper layered section overlying opaque, possibly true, basement. The junction between the two is presumed to correspond to the Hercynian orogenic phase.[2] The seismically opaque true basement has been sampled where it crops out on the seabed in the southwest of the Spur and comprises Middle Devonian (Late Eifelian to Early Givetian) or older shales and slates. Carboniferous rocks have also been sampled in the area. Dredgings on fault blocks south of Goban Spur have yielded granodiorites and various other Hercynian intrusives[1] and these are thought to locally cut the Palaeozoic sediments on the Spur.

The dipping succession within the tilted fault-blocks is thought to comprise an early Mesozoic pre-rift sequence of Triassic through Middle Jurassic age. Although this succession has not yet been sampled it is likely, by comparison with the succession in the Fastnet Basin, to consist of a Triassic continental clastic sequence overlain by a shaly and saliferous Keuper section, while the Jurassic may comprise shallow to moderately deep marine shales, marls and siltstones. The formation of the basement fault blocks in the Goban Spur may coincide with the major period of uplift at the end of the Middle Jurassic which produced igneous intrusion and faulting in the Fastnet Basin.

The lowest seismic layer unconformably overlying the faulted 'basement' has the main depocentres located in the northeast-trending interhorst grabens. This layer is characterized by small internal unconformities, minor folds, zones of fault-related disturbed bedding and rapid thickness variations.[1] Within the grabens, this lower sedimentary sequence thickens eastwards towards the main western boundary faults of the tilted fault blocks, and in a number of instances can be seen to drape over and onlap the basement ridges. Thicknesses up to 3 km are recorded for the layer, and it appears likely that the basement faults were still active during deposition of this lowest sedimentary sequence, which is probably predominantly Lower Cretaceous in age. A comparison with the Fastnet Basin suggests a Wealden age (Valanginian to Late Barremian) for the syn-rifting graben fill sequence. A fluvial and alluvial depositional environment, gradually deepening to a brackish and marginal marine setting could be expected in such a sequence. A correlation with the somewhat more distant Meriadzek Terrace might indicate a shallow marine Late Jurassic to Early Aptian sequence.

This syn-rift sequence in the Goban Spur is overlain, probably with a slight unconformity, by a relatively thin layer which rarely exceeds 1 km in thickness. By comparison with the Fastnet Basin it may represent a Greensand-equivalent sequence of Aptian to lowermost Cenomanian age comprising regressive marginal continental clastics giving way to marine inner sublittoral shales. A similar slight unconformity is often recorded at the base of this sequence in the Fastnet Basin. A comparison with the DSDP holes on the Meriadzek Terrace likewise suggests an Aptian through base Cenomanian sequence of predominantly shallow marine sediments. The sedimentation pattern and facies distribution of this sequence on the Goban Spur was, like the underlying sediments, again probably controlled by the tilted basement blocks and some slight faulting may have occurred during deposition of the layer. In general, the Goban Spur area remained a positive feature during deposition of the unit, with the probable consequential development of limestones, sandstones and shales in environments at or close to sea level.

The overlying sedimentary layer is markedly unconformable in the northern and southeastern parts of the Goban Spur. During deposition of this layer the main horst and graben topography which characterized the underlying two layers was obliterated in the northern part of the Goban Spur, although a strong northwest-southeast basement lineament is still evident at this level of the stratigraphy. The northern area was one of the main depocentres during deposition of this layer, with a major sediment source area lying immediately to the south. Sediment prograded and thickened northeastwards and also northwards into the deep southern part of the Porcupine Basin. A second depocentre at this level lay in the eastern part of the Goban Spur, with sediment probably prograding westwards from the adjacent Celtic Continental Shelf. A third main depocentre lay in the Porcupine Abyssal Plain, immediately west of the Goban Spur and of the oceanic–continental crustal boundary. However, throughout the Goban Spur area the thickness of this layer rarely exceeds 1 km and faulting is rare. Few of the basement horsts remained uncovered at the end of this phase of

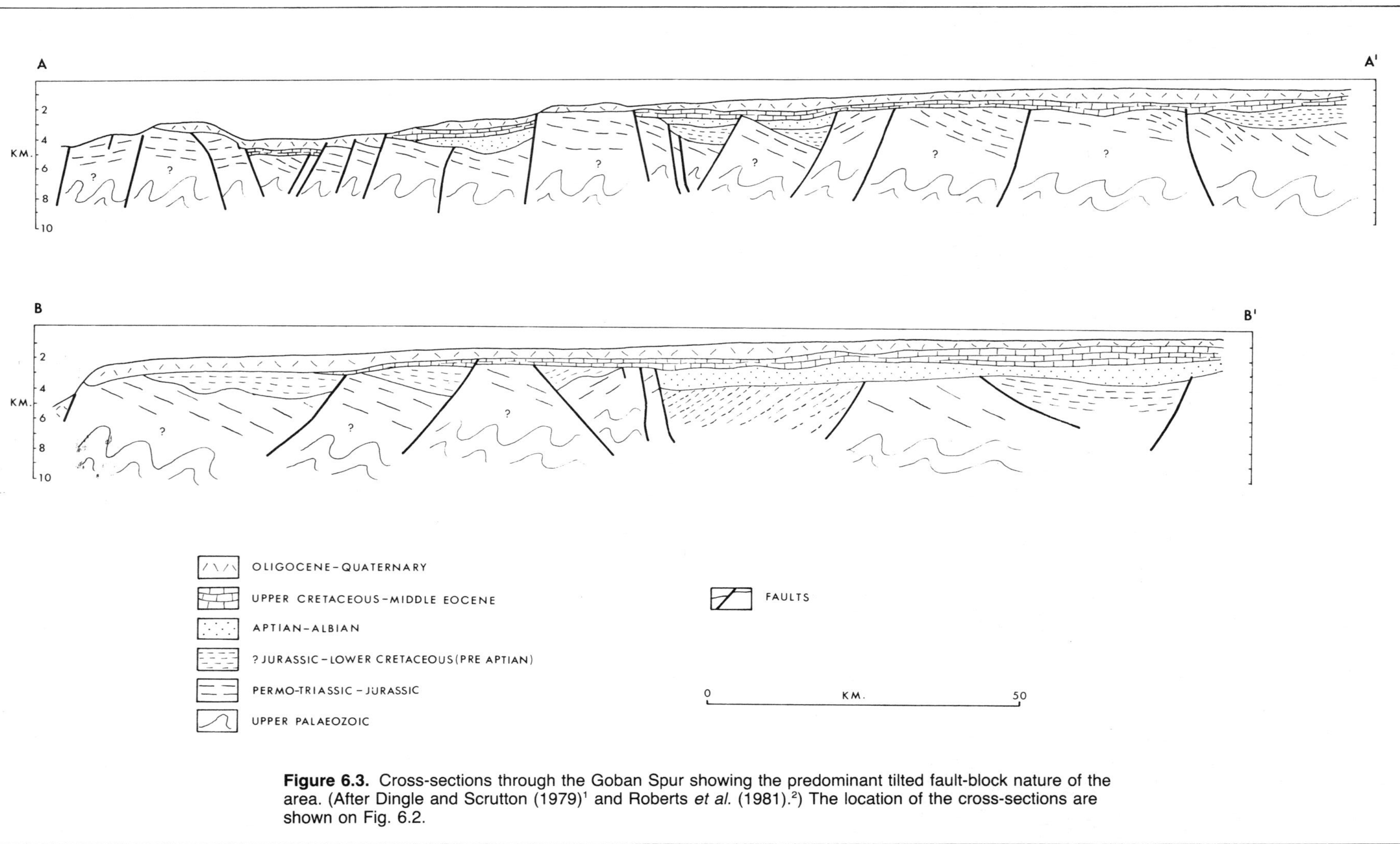

Figure 6.3. Cross-sections through the Goban Spur showing the predominant tilted fault-block nature of the area. (After Dingle and Scrutton (1979)[1] and Roberts *et al.* (1981).[2]) The location of the cross-sections are shown on Fig. 6.2.

deposition (Fig. 6.3). By comparison with the Fastnet Basin the layer is likely to be of Upper Cretaceous to Middle Eocene age and its base probably represents the mid-Cretaceous transgressive overstep at the base of a thick Chalk sequence. An outer shelf tranquil marine sedimentary regime is likely to have existed at the time. In support of this, some possible Upper Cretaceous Chalk samples have been recovered from the northwestern point of the Goban Spur. However, some probable deeper marine Upper Cretaceous samples were recovered from the southern Goban Spur, suggestive of at least localized deep marine depositional areas similar to those postulated for the equivalent layer encountered in the DSDP holes.

This sequence is overlain by a further complex layer containing a number of major and minor unconformities and showing substantial variations in thickness. It is separated from the underlying probable Chalk layer by an unconformity.

The sediments, like those of the underlying layer, are generally horizontally-stratified or very gently inclined and are suggestive of low energy marine sedimentation. However, some lateral variation appears in parts of the upper section of this layer, with again rather tranquil conditions persisting over most of the Goban Spur but slumping occurring in the vicinity of the scarps at the west of Goban Spur, while high energy scouring and moulding features were produced towards the northern end of the Spur. By analogy with the Fastnet Basin the sediments of this layer are likely to range from Middle Eocene to Quaternary in age. Few of the minor unconformities which occur in the sequence can be traced through the entire area but it is likely that a relatively major Miocene unconformity is widespread. A deepening marine sequence of thin limestones followed by a shaly succession with some coarser horizons might be anticipated.

Since Eocene time regional sedimentation rates in the Goban Spur area have generally remained relatively low, during a period when the main centre of sedimentation shifted northwards into the Porcupine Basin.

Hydrocarbon Potential

In view of the fact that no exploration wells have yet been drilled in the Goban Spur area, it is virtually impossible to draw a meaningful picture of the hydrocarbon potential of the area. However, following completion of their first major seismic survey, Esso are reported[8,9] as saying that the size of the structures seen in the area suggests that there is a possibility that fields containing 1 billion barrels of oil could exist in the Goban Spur. The published seismic sections[1,2] fail to show many such large structures, although it must be admitted that these data are not of the highest quality. In fact they indicate that the most likely traps would appear to be rather subtle unconformity traps, with the probability that most of the plays in the area will be clastic sediment wedges of Triassic to Lower Cretaceous age in a setting of tilted fault blocks. The reservoir potential is unknown, but in common with nearby basins significant thicknesses of porous and permeable sandstones might be expected at Triassic, Lower Cretaceous and possibly Middle and Lower Jurassic levels. Source rock potential is also unknown, although comparison with adjacent basins suggests possible source horizons at Jurassic and possibly Lower Cretaceous levels. However, as with the nearby Fastnet Basin there is the possibility that over much of the area the source rocks may not have been buried to sufficient depth to achieve thermal maturity, due to the area remaining relatively positive during much of its evolution.

In summary, while the Goban Spur area has not yet been drilled, it would seem likely that problems may exist with respect to source rock maturity and adequate structural traps. In addition some very significant technical problems could be anticipated while drilling in the formidable water depths and hostile environmental conditions in the area.

References

1. DINGLE, R. V. and SCRUTTON, R. A. 1979. Sedimentary succession and tectonic history of a marginal plateau (Goban Spur, Southwest of Ireland). *Mar. Geol.* **33**, 45–69.
2. ROBERTS, D. G., MASSON, D. G., MONTADERT, L. and de CHARPAL, O. 1981. Continental Margin from the Porcupine Seabight to the Armorican Marginal Basin: *in* Illing, L. V. and Hobson, G. D. (*Eds*). *Petroleum Geology of the Continental Shelf of North-West Europe.* Heyden & Son Ltd. London, 455–473,
3. BUNCE, E. T., CRAMPIN, S., HERSEY, J. B. and HILL, M. N. 1964. Seismic reflection observations on the continental boundary west of Bri-

tain. *J. geophys. Res.* **69**, 3853–3863.
4. SCRUTTON, R. A. 1979. Structure of the crust and upper mantle at Goban Spur—some implications for margin studies. *Tectonophysics* **59**, 201–215.
5. ROBINSON, K. W., SHANNON, P. M. and YOUNG, D. G. G. 1981. The Fastnet Basin: An Integrated Analysis: *in* Illing, L. V. and Hobson, G. D. (*Eds*). *Petroleum Geology of the Continental Shelf of North-West Europe.* Heyden & Son Ltd. London, 444–454.
6. MONTADERT, L., ROBERTS, D. G., AUFFRET, G. A., BOCK, W., DU PEUBLE, P.A., HAILWOOD, E.A., HARRISON, W., KAGAMI, H., LUMSDEN, D. N., MULLER, C., SCHNIFKER, D., THOMPSON, R. W., THOMPSON, T. L. and TIMOFEEV, P. P. 1977. Rifting and subsidence on passive continental margins in the North East Atlantic. *Nature* **268**, 305–309.
7. MONTADERT, L., ROBERTS, D. G., de CHARPEL, O. and GUENNOC, P. 1979. Rifting and subsidence of the northern continental margin of the Bay of Biscay: *in Intial Rep. Deep Sea Drill. Proj.* 48, US Govt. Printing Office, 1025–1060.
8. *FINANCIAL TIMES* NORTH SEA LETTER. 1980. Esso plays with high stakes in Goban Spur, 26th November, 272/7.
9. DAFTER, R. 1981. Irish Oil Prospects. Much depends on Porcupine. *Financial Times.* 12th January.

Chapter 7

Porcupine Basin

Introduction

The *Porcupine Seabight* is a large (320 × 240 km) and deep tongue-shaped area lying on the western margin of the Irish Shelf. It has an overall north-south trend. Water depths in the area are very formidable, deepening from about 350m in the north of Seabight to more than 1,700m in the south. The feature can be subdivided into a northern north–south-trending area of relatively shallower water and a trapezoidal-shaped northeast-trending area of very deep water which opens out into the *Porcupine Abyssal Plain*. The Seabight is bounded on three sides by relatively shallow platforms (Fig. 7.1). To the east lies the *Irish Mainland Shelf*, while the Seabight is bounded to the north by the east-west trending *Slyne Ridge* which connects the Galway coast to the bathymetrically-shallow *Porcupine Bank*. *Porcupine Ridge* extends southwards from Porcupine Bank to form the western boundary of the Seabight. The Seabight merges southwestwards into the Porcupine Abyssal Plain while to the south it rises, possibly across a slight barrier, to merge with the structurally higher but bathymetrically deep Goban Spur.

The bathymetry of this geologically complex area is a clear reflection of the underlying geology. The shallow features of Porcupine Bank, Porcupine Ridge and Slyne Ridge are all probably composed of ancient metamorphosed Precambrian and Caledonian rocks and deformed Upper Palaeozoic strata overlain by a thin layer of Tertiary sediments, with an occasional veneer of Mesozoic strata sandwiched between. The Seabight is the surface expression of an underlying deep sedimentary trough known as the *Porcupine Basin*. This basin, which has sharply defined fault-bounded eastern and western margins is infilled by great thicknesses of mainly Mesozoic and Tertiary sediments which exceed 5km in places.[1,2]

The Porcupine Basin appears to have a history of sedimentation which goes back at least to the early part of the Mesozoic Era. Like most of the other basins in the Irish Sea and west British offshore area, its evolution took place in the dominantly tensional structural regime which resulted from the early attempts at the opening of the Atlantic Ocean. In common with these other basins, the culminations of the compressive stresses of the Hercynian orogeny with their associated uplift of deformed strata in Late Carboniferous to Permian times was followed by a tensional rifting phase. This gave rise to the formation of fault-bounded grabens which were probably the sites of clastic sedimentation. A pattern of Early Mesozoic deposition in the Porcupine Basin broadly similar to that of Orphan Knoll (northeast of Newfoundland), Flemish Cap (off the Grand Banks of Canada) and possibly the Goban Spur, could be expected as all four basins were adjacent until approximately Late Cretaceous times (Fig. 1.3). At that time sea floor spreading commenced, firstly along the Newfoundland-Porcupine sector (possibly as early as 90 My BP) south of the Charlie Gibbs Fracture Zone (GFZ) and then later along the Labrador-Rockall sector (75 my BP) north of the GFZ.[3]

In Late Jurassic – Early Cretaceous times attempts at the opening of the North Atlantic Ocean were evident from the development of further rifting in the Rockall Trough. A short-lived parallel embryonic rift axis also appears to have developed in the Porcupine Basin. It may also have led, in pre-Upper Cretaceous times, to the clockwise rotation of the Porcupine Ridge, with the movement hinged about a location on the Slyne Ridge.[4]

During Coniacian times sea-floor spreading along the Newfoundland-Porcupine ridge axis, followed shortly afterwards in Campanian times by spreading along the Labrador-Rockall and Labrador-Greenland axes, resulted in the commencement of the separation of the Porcupine Basin and adjacent Rockall Trough from the eastern North American basins.[3] From then on similarities in sedimentation patterns between the Irish and North American basins began to decrease. Rapid subsidence and sedimentation during and after separation of the basins led to the development of a thick Tertiary section in the Porcupine Basin.

Geophysical Data

Seismic reflection profiling in the Porcupine Basin commenced in the early seventies and to date in excess of 40,000 km of seismic have been shot. Approximately half of this is now available for purchase by oil exploration companies from the Department of Industry and Energy in Dublin. In addition, a considerable amount of gravity and magnetic information, together with some further seismic reflection and rare seismic refraction data, has been accumulated and published by a number of academic scientific institutions (e.g. Institute of Oceanographic Studies, University of Edinburgh and the Marine Scientific Laboratories at University College of North Wales).[1,4–6] Sufficient geophysical information is therefore available to enable the relatively accurate delineation of the extent of the Porcupine Basin, together with a good approximation of the principal structural trends and sediment thicknesses and variations within the basin.

The gravity, magnetic and seismic reflection and refraction data all confirm the existence of a considerable thickness of Mesozoic sediments, sometimes in excess of 5 km, in the Porcupine Basin, which is clearly fault-bounded on both eastern and western flanks.[1,2] The geophysical data also show that the bounding shallow ridges (Irish Mainland Shelf, Slyne Ridge and Porcupine Ridge) are all characterized by continental crust approximately 30 km thick overlain by a generally thin veneer (less than 2 km in thickness) of Mesozoic to Recent sediments.

It has been argued by many workers that oceanic crust may be developed in a southwards-widening central tongue within the Porcupine Basin, while thin continental crust underlies the northern part of the basin. However, some new gravity and magnetic data, together with re-modelling and re-interpretation of existing information suggests that this oceanic crustal model is likely to be incorrect, and that the crustal model which best fits all the data is of an inactive and subsided rift of East African type underlain by continental crust. The latest ideas[7] suggest that continental crust, thinned to about 5 km and occasionally to as little as 2 km, exists underneath the entire Porcupine Basin. It is also likely that this thin crust is cut, especially in the central part of the basin, by a narrow zone of high density intrusions and/or extrusions of probable pre-Aptian age. However, further information, particularly by way of deep refraction data, is required before the 'thin continental crust' versus 'oceanic crust' argument is finally resolved.

A noticeable break in the gravity and magnetic data is seen across an east-southeast-trending line which crosses Porcupine Bank at approximately latitude 51°30′N and through the Porcupine Basin at approximately latitude 51°N, possibly continuing to the east-southeast along the boundary between the Fastnet Basin and the North Celtic Sea Graben. This break has been called the *Clare Lineament* and is thought to be a predecessor of the Charlie Gibbs Fracture Zone which lies adjacent to the west.[7]

The steep gravity gradients which culminate in a strong linear gravity high verify the location of the fault-bounded eastern limit of the basin and of the thinned continental crust.[6] The western margin shows a similar magnetic pattern indicating the shallow basement of the Porcupine Ridge. A pronounced gravity and magnetic anomaly over the eastern basement ridge at approximate latitude 53°N is interpreted as representing a large basic intrusive centre.[8] This is probable Tertiary in age and is likely comparable with centres of the Hebridean volcanic province. A series of dykes, many of which trend north-northwest, is also associated with the centre. The apparent extent of the centre and its associated intrusives suggests that some effects could be anticipated in the sediments along at least part of the eastern flank of the basin.

The seismic reflection data clearly show down-to-the-basin normal faults trending north-south parallel to the main eastern and western boundary faults. In addition, the basement appears to exhibit a remnant caledonoid trend, giving rise to northeast-trending faults.[8] The Porcupine Basin appears to be subdivided by a fault-bounded platform at approximate latitude 53°N into the Main Porcupine Basin and the smaller North Porcupine Basin (Fig. 7.1).

Seismic profiles show extensive thicknesses of Mesozoic to Recent sediments overlying an often block-faulted and tilted basement of probable Palaeozoic age.[6] Sediment thicknesses exceed 5 km in places. The seismic sections show us that a number of major seismostratigraphic units are developed throughout most of the basin. These are separated by well-defined reflectors representing regional unconformities or disconformities, while a number of smaller and less continuous unconformities are also seen throughout the area.

A possible change of sedimentation pattern is seen at approximate latitude 51°30′N, coinciding approximately with the so-called Clare Lineament. South of this latitude the main units are more or less horizontal, indicating uniform subsidence and sedimentation along the axis of the trough (Fig. 7.2).[6] Laterally prograding sets of strata occur in the vicinity of both margins of the basin. In the North Porcupine Basin the sedimentary succession is considerably thinner than further south. This is probably partly due to the Main Porcupine Basin having developed as a large trapdoor structure with greater downfaulting and resultant sedimentary infill occurring in the south than in the north. The two upper major seismic reflectors appear to merge together towards the north of the basin, suggesting the thinning of possible post-Cretaceous sediments northwards.

Stratigraphy

At the end of 1981 only nineteen wells had been drilled in the Porcupine Basin. The first well in these very deep waters off western Ireland was drilled by Shell in 1977 (Table 7.1). The wells, concentrated towards the northern end and the flank areas of the basin, have been drilled by a small number of major oil companies as follows: BP (5), Shell (4), Phillips (3), Deminex (2), Chevron (2), Elf (2), Gulf (1). At the present time all well information remains confidential and the stratigraphic picture outlined in this section is based mostly on information given in a number of recent publications[9,10] together with correlations with the better known sequences in the Fastnet Basin and the Eastern Canadian Basins (see Chapters 5 and 10).

Economic basement in the area is likely to comprise block-faulted mainly Upper Palaeozoic (Devonian and Carboniferous) sediments. The greater part of the basin lies north of the zone of intense Hercynian deformation and is therefore likely to show a lesser degree of metamorphism than time-equivalent basement rocks of the south coast basins (Fastnet and Celtic Sea). Upper Carboniferous Coal Measure strata crop out in the western part of Ireland in Counties Cork, Limerick and Clare, and similar coal-bearing sandstone-shale Westphalian sequences may be expected to occur in the Porcupine Basin.[9] Stephanian redbeds were also encountered in some wells drilled in the basin.[9]

The lowest Mesozoic unit comprises a very thick sequence of probable *Permo-Triassic* and *Jurassic* strata representing early rift-valley continental sediments overlain by shallow marine sequences.[9] By comparison with other areas the Permo-Triassic, if present, is likely to comprise a lower clastic sequence overlain by an upper saliferous succession. In places this Permo-Triassic sequence may reach thicknesses in excess of 2,000m.[9]

A Jurassic sequence of variable thickness is probably developed over much of the Porcupine Basin. The *Liassic* section, where present, is likely to comprise limestones and organic-rich shales with sandstones less frequently developed.[9] A marine depositional setting is envisaged for the Lower Jurassic.

Throughout much of Northwest Europe, and particularly the North Sea basins, the transition from the Liassic to the Middle Jurassic was marked by a major tectonic event, the Mid-Cimmerian rifting phase, which gave rise to extensive volcanic activity and erosion of older sediments.[9] By comparison, similar erosion in the Porcupine Basin could have removed much or all of the Liassic section from parts of the basin so that continental *Middle Jurassic* claystones and minor sandstones may unconformably overlie Liassic or older strata.[9] Middle Jurassic sandstones could be anticipated to have their best development towards the margins and the north of the Porcupine

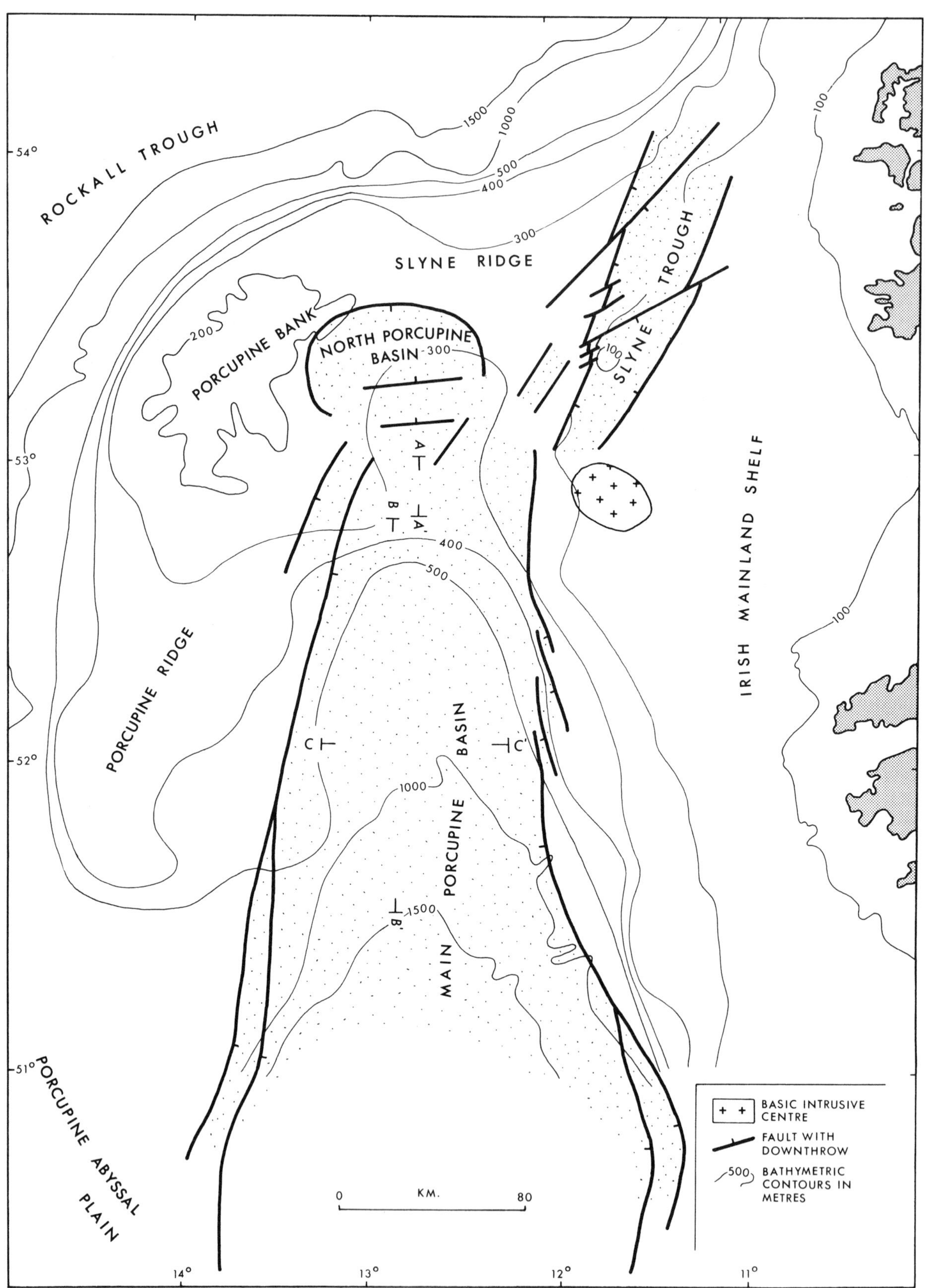

Figure 7.1. Geological sketch map for the Porcupine Basin (modified after Reeves, Robinson and Naylor (1978)[10]). showing the location of the main structural elements of the area.

Basin and may grade southwards into a marginal to fully marine sequence of shales and carbonates.[9]

Differential subsidence during latest Middle and *Upper Jurassic* times probably led to the development of continental to transitional and shallow marine sequences in the Porcupine Basin. These are likely to be predominantly shaly with some carbonates and clastics.[9] It is probable that sedimentation was continuous from Middle through Upper Jurassic times with no evidence of an unconformity at the base of the Upper Jurassic[11] similar to that seen in the Celtic Sea area. The upper surface of this unit is likely to correspond to the Late Cimmerian unconformity of latest Jurassic-earliest Cretaceous age.

The overlying sediments comprise a variable thickness (usually in excess of 750 m) of predominantly sandstones and shales.[9] The detailed stratigraphy of the *Cretaceous* and *Tertiary* in the deep southern part of the Porcupine Basin is poorly understood or documented. Therefore the post-Jurassic sequence is illustrated by the results of work in the northern part of the Main Porcupine Basin where seismic stratigraphic analysis[10] has revealed a relatively detailed post-Late Cimmerian sedimentation history. Frequent fluctuation in sea level is reflected in the transgression and regression of interfingering sandstone-, shale-, limestone- and chalk-dominated facies. In this northern part of the Main Porcupine Basin the undulatory Late Cimmerian unconformity surface is overlain by a northwards-onlapping sequence of deep sea sandstone-dominated deposits. These in turn are overlain by a southward-prograding series of shale-dominant bottomset and foreset sediments, with accompanying sand-prone topset deposits. A relative fall in sea-level resulted in the development of an emergent platform with accompanying slope deposits comprising a lower shale series and an overlying northward-onlapping sandstone-dominant sequence resulting from the erosion of the emergent platform. A relative rise in sea level produced a sequence of offshore sandstone bars towards the north (Fig. 7.2).

This was then followed by a further relative rise in sea level, marking the onset of the Upper Cretaceous, with the development of a northward-thinning and onlapping outer shelf to slope sequence of carbonates. Something of the order of 500 m of Upper Cretaceous is thought to be preserved near the margins of the basin, while in excess of 1,000 m probably occurs towards the central axis.[9] A slight relative drop in sea level in Palaeocene time produced a southward-prograding carbonate shelf displaying a slight downlapping relationship with the underlying Upper Cretaceous Chalk (Fig. 7.2).

The Upper Palaeocene to Lower Eocene in the Main Porcupine Basin is marked by a complex series of prograding deltaic sequences. Some of these built up from north to south along the axis of the basin, while others, probably fault-related, prograded eastwards from the western margin. In general, topsets tend to contain sandstones while foresets and bottomsets are usually shale-dominant. This period of deltaic progradation occurred during a period of relative rise of sea level, as suggested by the northwards shift of coastal onlapping of the deltaic system, with the sea level rise being largely balanced by an increased clastic supply.

The Upper Eocene is marked by a further relative rise in sea level accompanied by the development of offshore bar sandstones. These are unconformably overlain by late Upper Eocene to Quaternary deposits comprising deep sea sands and shales. Channel sandstones are frequently recognized on seismic profiles, while shale-prone levee deposits are also common.

A substantial thickness of Tertiary sediments is preserved throughout the Porcupine Basin. This reaches a maximum in the central and southern areas where upwards of 3 km of limestones, sandstones and shales are preserved. There is thinning towards the north and to the basin edges where in the order of 500m of such sediments might be expected.[9]

Hydrocarbon Potential

Of the small number of exploration wells drilled in the Porcupine Basin to date (Table 7.1), quite a few are reported to have encountered oil shows while four wells (Aran/BP 26/28-1 and 26/28-2, Phillips 35/8-1 and 35/8-2) have recorded oil flows of 5,589 b/d, 1,490 b/d, 730 b/d and 925 b/d, respectively. The quality of the oil recorded from the basin is good, and in general appears to be a sulphur-free sweet light crude, with API gravities in the range 30–41°. Little is known about the size of the oil-bearing structures, but the drilling results to date suggest that they are likely to be complex and small or moderate in size by world standards. So far, however, no commercial hydrocarbon prospects have been fully delineated

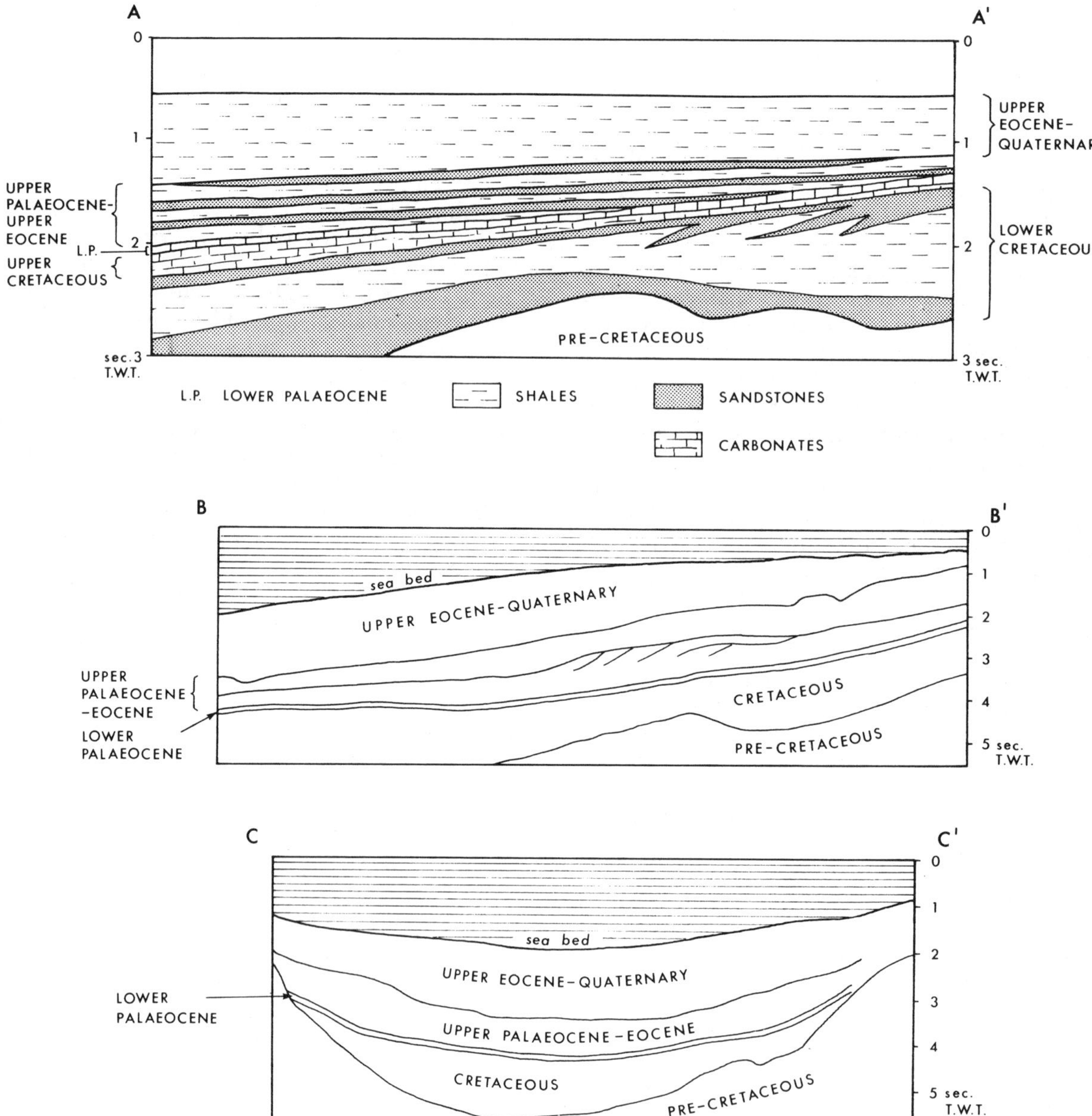

Figure 7.2. Cross-sections based on seismic reflection profiles through the Main Porcupine Basin. (Modified from Louis and Mermey (1979).[8]) The location of the cross-sections is shown on Fig. 7.1.

or appraised in the Porcupine Basin. The prospect on which most drilling, and consequent press speculation, has taken place is in Aran/BP block 26/28 which lies in some 370 m of water approximately 150 km west of the Irish mainland. To date four wells have been completed, with the first two wells flowing oil on test and the second two being dry holes. The prospect, although it contains good quality Jurassic reservoir sandstones,[12,13] appears to be structurally very complex and contains both oil-bearing and water-wet individual fault segments. Consequently it is thought to be too early to give any accurate estimate of the reserves contained in the structure. It is generally considered that further exploration work, including drilling, will be required before a decision can be made on whether or not oil exists in commercial quantities in the area.

TABLE 7.1 Wells drilled in the Porcupine Basin

Well name	*Area*	*Latitude Longitude*	*Spud date Rig release date*	*R.T. elevation Water depth*	*Total depth Well status*
Shell/Agip 35/13-1	Main Porcupine Basin	52°35′51.980″N 12°25′11.820″W	7.04.′77 2.11.′77	24m 482m	4,405m Oil shows P. & A.
Deminex 34/15-1	Main Porcupine Basin	52°33′31.408″N 13°07′22.020″W	4.06.′77 24.09.′77	23m 482m	4,446m Oil shows P. & A.
Shell/Agip 35/29-1	Main Porcupine Basin	52°06′50.200″N 12°18′28.200″W	3.04.′78 29.05.′78	24m 804m	2,900m P. & A.
Elf 35/2-1	Main Porcupine Basin	52°58′32.700″N 12°36′47.500″W	7.04.′78 2.10.′78	13m 391m	3,995m Oil shows P. & A.
Phillips 35/8-1	Main Porcupine Basin	52°43′01.783″N 12°25′06.456″W	4.05.′78 3.10.′78	26m 430m	4,419m Oil flows (730 b/d) P. & A.
Shell/Agip 34/19-1	Main Porcupine Basin	52°26′15.230″N 13°20′53.070″W	31.05.′78 10.08.′78	24m 426m	3,208m P. & A.
Aran/BP 26/22-1	N. Porcupine Basin	53°13′36.780″N 12°42′24.370″W	12.07.′78 22.07.′78 (02.10.′78)	26m 343m	805m Not completed for technical reasons P & A.
Aran/BP 26/22-1A	N. Porcupine Basin	53°13′37.090″N 12°42′48.890″W	26.07.′78 28.09.′78	26m 346m	2,573m P. & A.
Deminex 35/6-1	Main Porcupine Basin	52°42′16.700″N 12°57′37.300″W	13.08.′78 19.11.′78	24m 486m	3,959m Oil shows P. & A.
Chevron 36/16-1	Main Porcupine Basin	52°28′37.166″N 11°56′18.288″W	24.05.′79 re-spudded 29.05.′79 17.08.′79	30m 250m	2,745m P. & A.
Aran/BP 26/28-1	Main Porcupine Basin	53°02′29.182″N 12°33′31.729″W	12.06.′79 30.08.′79	28m 374m	3,315m Oil flows (5,589 b/d) P. & A.
Gulf 26/21-1	N. Porcupine Basin	53°11′15.522″N 12°49′46.466″W	7.09.′79 12.10.′79	26m 338m	2,176m P. & A.
Phillips 35/15-1	Main Porcupine Basin	52°34′54.299″N 12°06′43.803″W	24.06.′80 19.10.′80	30m 308m	3,687m P. & A.
Aran/BP 26/28-2	Main Porcupine Basin	53°02′26.290″N 12°34′49.240″W	11.07.′80 24.09.′80	26m 378m	2,697m Oil flows (1,440 b/d) P. & A.
Elf 34/5-1	Main Porcupine Basin	52°59′27.472″N 13°10′43.940″W	18.07.′80 re-spudded 22.07.′80 19.08.′80	25m 289m	1,448m P. & A.
Chevron 26/27-1	Main Porcupine Basin	53°05′16.290″N 12°43′59.590″W	6.04.′81 13.04.′81	26m 383m	666m Not completed for technical reasons P. & A.
Chevron 26/27-1A	Main Porcupine Basin	53°05′16.290″N 12°43′59.590″W	14.04.′81 22.04.′81	26m 383m	666m Not completed for technical reasons P. & A.
Chevron 26/27-1B	Main Porcupine Basin	53°05′17.040″N 12°43′59.400″W	22.04.′81 19.06.′81	26m 383m	2,815m P. & A.

TABLE 7.1 – continued

Well name	*Area*	*Latitude Longitude*	*Spud date Rig release date*	*R.T. elevation Water depth*	*Total depth Well status*
Phillips 35/8-2	Main Porcupine Basin	52°46′30.198″N 17°27′54.617″W	19.04.'81 23.10.'81	26m 422m	4,633m Oil and gas flows (925 b/d, 4.853 MMcfd) P. & A.
Aran/BP 26/28-3	Main Porcupine Basin	53°04′15.000″N 12°31′26.650″W	26.05.'81 15.07.'81	26m 364m	2,588m P. & A.
Shell 26/26-1	Main Porcupine Basin	53°04′44.691″N 12°51′57.612″W	2.06.'81 30.06.'81	12m 363m	1,256m Oil shows P. & A.
Aran/BP 26/28-4	Main Porcupine Basin	53°04′26.050″N 12°27′42.316″W	22.07.'81 8.08.'81	26m 350m	761m Not completed for technical reasons P. & A.
Aran/BP 26/28-4A	Main Porcupine Basin	53°04′24.649″N 12°27′41.982″W	9.08.'81 8.10.'81	26m 350m	2,418m P. & A.

It is probable that a number of source and reservoir horizons occur within the Porcupine Basin. If, as is thought likely, Carboniferous coal-bearing strata extend westwards from the Irish mainland as far as the Porcupine Basin, they are likely to have the potential for the generation of gas. The general area is far enough north for the effects of Hercynian deformation to be relatively slight. In addition, Carboniferous sandstones could also have some, albeit limited, reservoir potential. Permo-Triassic strata, if present, are also likely to be potential reservoir exploration targets, but no source rock potential is expected at this level. By comparison with many Mesozoic basins throughout Northwest Europe, possible the best source and reservoir horizons in the Porcupine Basin will occur at Middle and Upper Jurassic level. Organic-rich shales of this age, buried to sufficient depth to generate hydrocarbons, might be expected. Reservoir rocks, with good flow characteristics are known to occur at Jurassic level.[12,13] In BP 26/28-1 three approximately 10m thick intervals of Jurassic sandstones flowed oil, while two different Jurassic intervals flowed oil on test in BP 26/28-2. The Jurassic is therefore likely to be a major exploration target in the basin. The Lower Cretaceous must also be considered to be of interest since it contains both sandstones and shales and therefore, if it has been buried deeply enough to generate hydrocarbons, it is likely to have the potential to act as both source and reservoir. The Tertiary, although it probably contains appreciable quantities of reservoir quality sandstones, is likely to be thermally immature throughout much of the basin and may lack adequate capping sediments and sealing traps.

In summary therefore, the relatively virgin large Porcupine Basin with its thick Mesozoic sedimentary succession appears at this early stage of exploration to have good overall prospectivity. However, water depths and hostile environmental conditions are likely to continue, for some time at least, to remain as major obstacles to both hydrocarbon exploration and eventual production.

References

1. BAILEY, R. J. 1975. The geology of the Irish continental margin and some comparisons with offshore eastern Canada: *in* Yorath C. J. (*Ed.*). *Canada's Continental Margins and Offshore Petroleum Exploration. Can. Soc. Pet. Geol. Mem.* **4**, 313–340.
2. REEVES, T. J., ROBINSON, K. W. and NAYLOR, D. 1978. Ireland's Offshore Geology. *Irish Offshore Review* **1**, 25–28.
3. SCRUTTON, R. A. 1981. North Atlantic regional tectonics and continental reconstructions: *in* Lecture Notes accompanying one day

meeting on *'Geology and Geophysics of the U.K. and Irish continental shelf – Goban Spur to Rockall Bank'*, pp. A1–A14.

4. BAILEY, R. J., BUCKLEY, J. S. and CLARKE, R. H. 1971. A model for the evolution of the Irish continental margin. *Earth planet. Sci. Lett.* **13**, 79–84.
5. ROBERTS, D. G., MASSON, D. G., MONTADERT, L. and de CHARPAL, O. 1981. Continental Margin from the Porcupine Seabight to the American Marginal Basin: *in* Illing, L. V. and Hobson, G. D. *(Eds) Petroleum Geology of the Continental Shelf of North-West Europe*. Heyden & Son Ltd. London, 455–473.
6. BAILEY, R. J. 1979. The Continental Margin from 50N to 57N: its Geology and Development: *in* Banner, F. T., Collins, M. B. and Massie, K. S. *(Eds). The North-West European Shelf Seas: the Sea Bed and the Sea in Motion. 1. Geology and Sedimentology*. Elsevier, Amsterdam, 11–24.
7. MEGSON, J. 1981. Crustal structure of the continental margin west of the U.K. and Ireland: *in* Lecture notes accompanying one day meeting on *'Geology and Geophysics of the U.K. and Irish continental shelf – Goban Spur to Rockall Bank'*. pp. B1–B22.
8. RIDDIHOUGH, R. P. and MAX, M. D. 1976. A geological framework for the continental margin west of Ireland. *Geol. J.* **11**, 109–118.
9. ZIEGLER, P. A. 1981. Evolution of Sedimentary Basins in North-West Europe: *in* Illing, L. V. and Hobson G. D. *(Eds). The Geology of the Continental Shelf of North-West Europe*. Heyden & Son Ltd. London, 3–39.
10. LOUIS, P. R. and MERMEY, P. 1979. An example of seismic stratigraphy. The Porcupine Basin-Western Ireland. Paper presented at *49th Annual International Meeting of the Society of Exploration Geophysicists*, New Orleans.
11. CASTON, V. N. D., DEARNLEY, R., HARRISON, R. K., RUNDLE, C. C. and STYLES, M. T. 1981. Olivine dolerite intrusions in the Fastnet Basin. *J. geol. Soc. London* **138**, 31–46.
12. LACEY, J. R. 1980. Petroleum Consultants' Report: *in Aran Energy Rights Issue document.*
13. LACEY, J. R. 1981. Petroleum Consultants' Report: *in Bula Resources Rights Issue document.*

Chapter 8

Northwest Irish Offshore Basins

Introduction

A chain of partly connected, poorly understood and probably relatively complex, narrow sedimentary troughs extends northeastwards from the neck of the Porcupine Basin. These basins lie some 50 km offshore from the northwesterly landfalls of the Irish mainland (Fig. 8.1) and are described in this chapter under the collective title of Northwest Irish Basins. Unlike the large and adjacent Porcupine Basin, which shows a predominant north-south trend imposed by early Mesozoic rifting, these basins all show a marked caledonoid northeast-southwest trend. The main southern part of this sedimentary complex is called the *Slyne Trough* and it abuts against the northeastern end of the main Porcupine Basin. A further small area of Tertiary, and possibly Mesozoic and Upper Palaeozoic, sedimentary infill runs subparallel to the southwestern end of the Slyne Trough (Fig. 8.1). The main Slyne Trough is cut and offset by a series of caledonoid-trending faults which are at an acute angle to the trough margins. The faults produce an *en echelon* pattern to the trough as it crosses the east-west trending basement feature of the *Slyne Ridge*.

North of the Slyne Ridge but *en echelon* to, and possibly structurally distinct from, the Slyne Trough lies the *Erris Trough*. A partial connection may exist with the massive Rockall Trough to the west. The Erris Trough in turn passes northeastwards into a structurally complex area which widens northwards into the main *Donegal Basin*. This basin, which appears to be separated by a narrow ridge from the Rockall Trough, is probably terminated at its northeastern extremity by a northwest-southeast fault (Fig. 8.1).

Water depths in this general area are quite varied. The Slyne Trough is the shallowest of the three main basins with water depths between 150 m and 250 m, while the Erris Trough has the deepest water in the area ranging from 150 m up to 1,500 m. In the Donegal Basin water depths are usually of the order of 100–300 m, but in places exceed 500 m.

As in the adjacent major Porcupine and Rockall Basins, the Northwest Irish Basins are likely to have a depositional history extending back to Late Palaeozoic time. The sedimentary infill and tectonic episodes may be expected to reflect Early Mesozoic crustal rifting during the abortive attempts of the North Atlantic Ocean to open northwards through the Rockall and Porcupine Basins. However, unlike these two major basins there is no evidence in the narrow northwestern grabenal basins of significant thinning of continental crust beneath the Mesozoic sedimentary cover.

Similarities in terms of depositional environments and facies types can be expected between the Northwest Basins and the Porcupine Basin to the south and also, albeit to a lesser extent, with the West Shetlands and Hebridean Basins which lie roughly along strike to the northeast.

Geophysical Data

Although several thousand kilometres of deep seismic reflection data have been acquired by oil exploration companies since the commencement of exploration of the area in the early 1970s, little has been published concerning these small and relatively deep water sedimentary basins. However, a certain amount of gravity and magnetic data, together with shallow reflection seismic profiles, have been acquired and published by academic institutions, with the most prolific of these being the Marine Science Laboratories of University College, North Wales.[1, 2] Most of this work has been concentrated on the Slyne Trough, with relatively little attention given to the Erris Trough or the Donegal Basin.

The geophysical data indicate that the Slyne Trough is a narrow (about 25 km in width) and generally parallel-sided, half grabenal feature. Approximately 2.5 km of Mesozoic and Tertiary sediments fill the trough[3, 4] and these show a predominant westerly dip towards the major westerly boundary fault. The Erris Trough and Donegal Basin are also likely to be essentially half-grabenal structures containing something of the order of 2 km of Mesozoic and Tertiary strata.[3]

Gravity and magnetic information suggests that the basement rocks underlying the basins are likely to be similar to the basement rocks of the Irish mainland. The total field magnetic anomaly pattern of the area reveals a number of basement features throughout the Slyne, Erris and Donegal Basins. In the north of the Donegal Basin, a complex positive anomaly is interpreted in terms of a buried ridge of metamorphic basement rocks, an interpretation which is also consistent with the available gravity data.[1, 4] This anomaly broadens northwards and diminishes, probably implying deeper burial of the basement. A smaller and again complex positive anomaly occurs further south in the Donegal Basin (at approximately 55°N) and is likewise interpreted as a buried basement ridge. In common with the larger northern ridge it is also fault-bounded. The gravity information suggests that, at depth, a further fault-bounded basement ridge separates the Slyne Trough from the adjacent northeastern sector of the Porcupine Basin.[4] At higher levels the relationships between the two basins are somewhat more obscure and there may even have been some direct sedimentary interconnection from the Slyne Trough into the Porcupine Basin during Early Mesozoic time. The magnetic anomaly pattern immediately west of the southwestern end of the Slyne Trough tentatively suggests the presence of a regional dyke swarm with a north-northeasterly trend.[4] This would appear likely in view of the known presence of a major Tertiary igneous centre adjacent to the northeastern margin of the Porcupine Basin[5] and of the Tertiary basic intrusives of the Hebrides Terrace Seamount northwest of the Donegal Basin. In the light of such extensive Tertiary igneous activity in the vicinity many other instances of similar intrusions could also be expected with the basins.

Two predominant fault directions occur within the basins. The major trend has a northeast-southwest caledonoid orientation. This is probably basement-controlled and appears to parallel, and indeed may be related to, the similar-trending Great Glen Fault system. The second direction is approximately east-west, and may likewise represent a reactivated older basement-controlled fault system.

Stratigraphy

Little is known, and even less has been published, of the stratigraphy or sedimentology concerning the Northwest Irish Basins. Only four wells have been drilled in the area to date (early 1982). Two of these (Texaco 13/3-1 and Amoco 12/13-1A) were located in the Donegal Basin, one (Amoco 19/5-1) was in the Erris Trough and one (Elf Aquitaine 27/13-1) lay in the Slyne Trough (Table 8.1). The information from all four wells remains confidential. The stratigraphic picture outlined below is based on very limited published data, including comparisons and extrapolations from the West Shetlands and Hebrides areas along strike to the northeast and the Porcupine Basin to the southwest, together with extrapolation from the two DSDP wells drilled to the west of the Rockall Bank.[6]

It is likely that the sedimentary fill in the basins is Late Palaeozoic to Tertiary in age. The Elf-Aquitaine 27/13-1 well is known to have encountered a Mesozoic section. The total sediment thickness in the Slyne Trough is probably of the order of 2.5 km,[3] while something in excess of 2 km of post-Palaeozoic sediments are probably present in the Erris Trough and Donegal Basin.[3]

Several wells in the offshore West Shetlands area have encountered a *Devono-Carboniferous* succession resting on Precambrian (Gren-

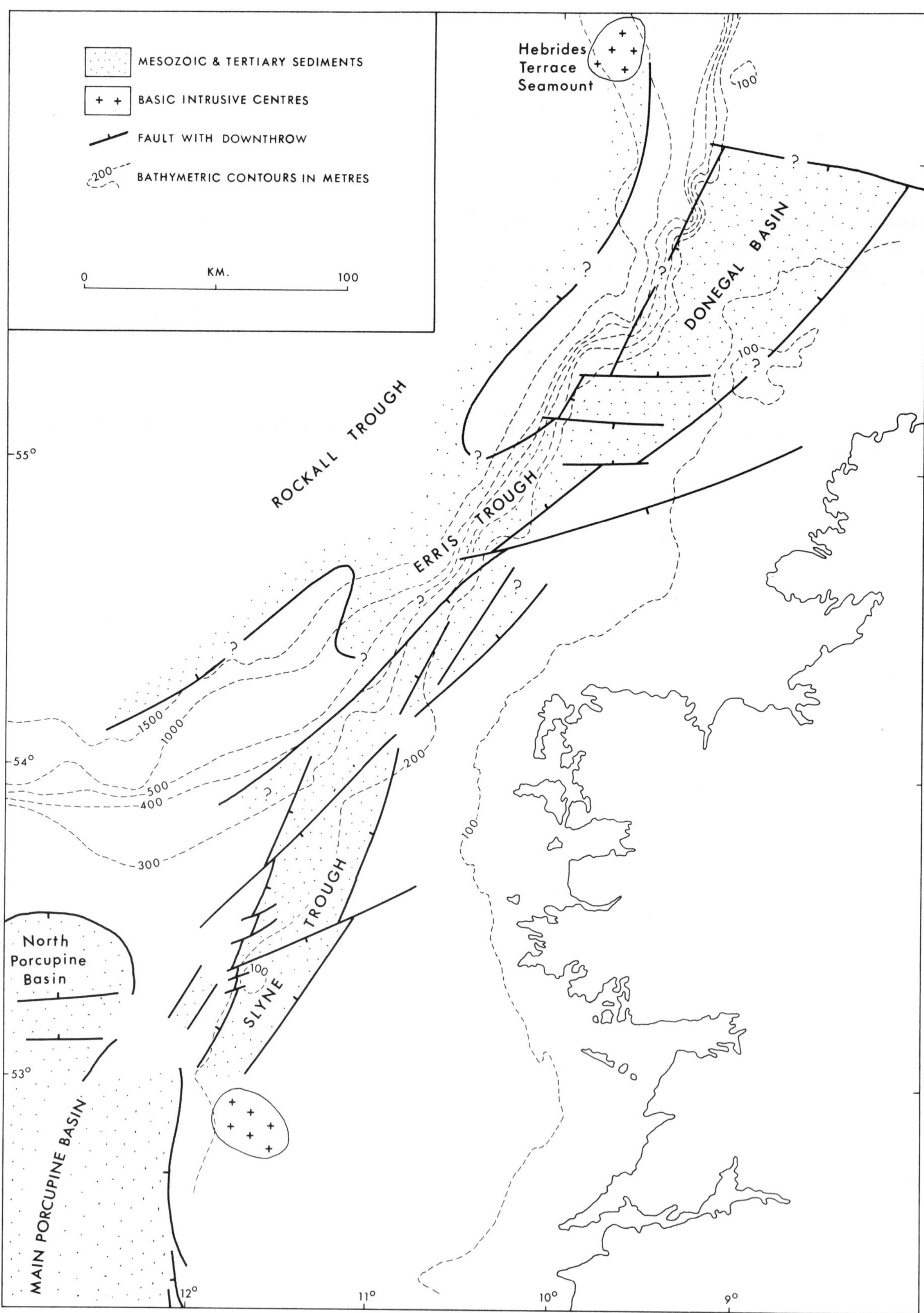

Figure 8.1. Geological map showing the location and main structural features of the Northwest Irish Offshore Basins. (Modified from Reeves, Robinson and Naylor (1975)[6] and Bailey (1979).[1])

villian) basement (see Chapter 11). A general fluvial and alluvial setting is envisaged with deposition of sandstone- and mudstone-dominant intervals over an irregular Precambrian land surface. A similar succession occurs in the onshore Northwest Carboniferous Basin and may be expected to extend offshore in the Slyne-Erris-Donegal area. Upper Carboniferous strata, sometimes coal-bearing, crop out onshore on the western and northwestern Irish mainland. These could therefore be expected to continue westwards into at least part of the area now occupied by the Northwest Irish Basins.

The Devono-Carboniferous sequence is overlain in the West Shetlands area by a red-bed succession of sandstone-dominant *Permo-Triassic* sediments which in places exceed 1 km in thickness. An alluvial depositional regime is envisaged. Onshore in the adjacent Inner Hebrides up to 300 m of Triassic sediments occur. These comprise conglomerates and sandstones overlain by clay-and marl-dominant sequences for which a continental environment is envisaged. In view of this, and of the general occurrence of early syn-rift Permo-Triassic continental sediments in most offshore Irish sedimentary basins, the occurrence of a lower sandy and upper shaly/saliferous Permo-Triassic sequence in the Slyne, Erris and Donegal Basins is thought likely.

Although *Lower Jurassic* sediments are generally absent from the West Shetland Basins, up to 1 km of such sediments are preserved beneath the Tertiary extrusives of the Inner Hebrides. The Lower Liassic section comprises a marine limestone succession, representing the widespread Lower Jurassic marine transgression, overlain by a thick sequence of dark shales. These in turn are succeeded by Middle Liassic sandstones and Upper Liassic shales. This is similar in many respects to the Liassic sequence encountered in the Fastnet Basin and to that proposed for the Porcupine Basin. A succession bearing some similarities to this might therefore be anticipated throughout most of the Northwest Irish Basins.

The *Middle Jurassic* is present in the West Shetlands and Inner Hebrides areas and in both cases marginal marine and non-marine sandstones tend to be the predominant lithology. Overall, thicknesses of the Middle Jurassic are rather small and are of the order of up to 100 m in the West Shetlands areas.

Upper Jurassic sediments, predominantly comprising an organic-rich marine shale sequence of Oxfordian to Kimmeridgian age, occur in both the West Shetlands and Inner Hebrides areas. However, it is thought unlikely that any Upper Jurassic strata are preserved in at least the northern part of the Northwest Irish Basins (Fig. 14.4).

By comparison again with most of the adjacent basins the main Cimmerian tectonic phase may have produced a major unconformity at the base of the *Cretaceous* in the Slyne-Erris-Donegal Basins. The *Lower Cretaceous* in the West Shetland and Inner Hebrides areas is composed of shelf and deeper marine sandstones, shales and limestones. Very variable thicknesses occur, ranging from less than 100 m to more than 1,000 m. Similar facies and variable sediment thicknesses might therefore be anticipated in the Northwest Irish Offshore Basins, with the Lower Cretaceous either very thin or absent throughout much of the area.

The Lower Cretaceous marine sequence of the Inner Hebrides is overlain by *Upper Cretaceous* Chalk whilst in the West Shetlands area open marine carbonates of similar age are widespread. Upper Cretaceous Chalk is also indicated on seismic stratigraphy evidence from the northern part of the Main Porcupine Basin. Its deposition through the Northwest Irish Basins is therefore thought to be highly likely, although, as in other basins, it may be thin or even absent by subsequent erosion in places.

A major unconformity at the base of the *Tertiary* is recorded in a number of the Irish basins, in addition to the Hebrides and West Shetlands areas. In the latter areas the Tertiary ranges from Palaeocene through to Plio-Pleistocene with an important unconformity at the top of the Miocene. Lithologies comprise extensive Late Palaeocene to Early Eocene eruptives overlain by marine mudstones, sandstones and limestones. No geophysical evidence exists in the Northwest Irish Basins to suggest the presence of major amounts of similar Early Tertiary extrusives. It is likely that an extensive and substantial thickness, perhaps as much as 1,000 m, of Tertiary sediments was deposited in the area. The limited available seismic coverage indicates the presence of a number of Tertiary unconformities, some of which have been compared and equated with similar seismic breaks in the adjacent Rockall Trough. It has been suggested that one of these may mark an early Oligocene hiatus. If so it is possible that the onset of Hebridean volcanism in western Scotland was marked by associated crustal instability and related uplift and erosion without the occurrence of massive erup-

TABLE 8.1 Wells drilled in the Northwest Irish Offshore Basins

Well name	*Area*	*Latitude Longitude*	*Spud date Rig release date*	*R.T. elevation Water depth*	*Total depth Well status*
Texaco 13/13-1	Donegal Basin	55°53′18.678″N 08°30′08.214″W	27.05.′78 25.06.′78	26m 131m	1,366m P. & A.
Amoco 19/5-1	Erris Trough	54°57′42.213″N 10°02′00.109″W	28.06.′78 13.08.′78	27m 116m	2,584m P. & A.
Amoco 12/13-1	Donegal Basin	55°37′03.010″N 09°30′39.090″W	16.06.′79 08.07.′79 (11.09.′79)	12m 480m	872m Not completed for technical reasons P. & A.
Amoco 12/13-1A	Donegal Basin	55°37′03.650″N 09°30′38.250″W	17.07.′79 9.09.′79	12m 480m	2,869m P. & A.
Elf 27/13-1	Slyne Trough	53°33′26.700″N 11°26′37.000″W	17.10.′81 (re-spudded 22.10.′81) 24.01.′82	24m 193m	

tive activity in the Slyne-Erris-Donegal Basins.

Hydrocarbon Potential

The four wells which had been drilled and completed by early 1982 in the Northwest Irish Offshore Basins were plugged and abandoned as dry holes without encountering significant hydrocarbon shows. A number of blocks have been relinquished in the area and no new licences have recently been granted. It would therefore appear that the area is currently viewed by the majority of the oil industry as not having the same hydrocarbon potential as a number of the other major Irish basins.

However, with only four wells completed in the Northwest Irish Basins, the general area must still be regarded as having some reasonable prospectivity. The known major faults allow for the juxtapositioning of potential source and reservoir rocks while the indicated tectonic events and unconformities may also provide trapping mechanism. There is no published data on the quality of source and reservoir rocks in the region. However, comparisons with the West Scottish Basins and to a lesser extent the Porcupine Basin suggest possible reservoir development at Carboniferous, Permo-Triassic, Lower Cretaceous and possible Middle Liassic and Middle Jurassic levels, where these are present. It is unlikely that Upper Cretaceous or Tertiary horizons, if present, will be prospective due to inadequate traps at these levels. Source potential, again by analogy with the West Scottish Basins, may exist in Liassic shales, and in Middle Jurassic shales if these occur in the basins. In the Inner Hebrides the Bathonian Dun Caan Oilshale unit has rich hydrocarbon source potential. If Lower Cretaceous shales are present they may represent a possible potential source horizon, although it is doubtful if they will have reached the oil generation threshold, except in local areas, due to lack of burial. If Upper Carboniferous coal-bearing strata, comparable to those onshore on the western Irish mainland, underlie the basins they may constitute a potential gas source which is likely to have been buried sufficiently deeply to have reached thermal maturity.

In summary, it is likely that a number of potential source and reservoir horizons may be present in the Slyne, Erris and Donegal Basins, while trapping mechanisms are also expected.

References

1. BAILEY, R. J. 1979. The Continental Margin from 50N to 57N: its Geology and Development: *in* Banner, F. T., Collins, M. B. and Massie K. S. (*Eds*). *The North-West European Shelf Seas: the Sea Bed and the Sea in Motion. 1. Geology and Sedimentology.* Elsevier, Amsterdam, 11–24.
2. BAILEY, R. J., JACKSON, P. D. and BENNELL, J. D. 1977. Marine geology of Slyne Ridge. *J. geol. Soc. London* **133**, 165–172.

3. REEVES, T. J., ROBINSON, K. W. and NAYLOR, D. 1978. Ireland's Offshore Geology. *Irish Offshore Review* **1**, 25–28.
4. BAILEY, R. J. 1975. The geology of the Irish continental margin and some comparisons with offshore eastern Canada: *in* Yorath, C. J. (*Ed.*). *Canada's Continental Margins and Offshore Petroleum Exploration*. Can Soc. Pet. Geol. Mem. **4**, 313–340.
5. RIDDIHOUGH, R. P. and MAX, M. D. 1976. A Geological Framework for the Continental Margin to the west of Ireland. *Geol. J.* **11**, 109–120.
6. MONTADERT, L., ROBERTS, D. G., AUFFRET, G. A., BOCK, W. D., DUPEUBLE, P. A., HAILWOOD, E. A., HARRISON, W. E., KAGAMI, H., LUMSDEN, D. N., MULLER, C. M., THOMPSON, R. W., THOMPSON, T. L. and TIMOFEEV, P. P. 1979. *Initial Reports of the Deep Sea Drilling Project.* Vol. XLVIII, Washington, U.S. Govt. Printing Office. 1183 pp.

Chapter 9

Rockall Plateau and Trough

Introduction

The foot of the continental slope marking the western margin of Europe lies far to the west of Ireland and Scotland (Fig. 1.1). The area within the margin shows varied physiography with the deeper water areas of the Porcupine Seabight, Rockall Trough and Hatton-Rockall Basin separating areas or banks covered by shallower water. The topographic features of the westernmost part of the continental margin bordering the deeper Atlantic are normally grouped under the names *Rockall Plateau* (Fig. 9.1) and *Porcupine Bank*.

Rockall Plateau is a relatively shallow water region, covering almost 25,000 square kilometres within the 2,000 m isobath. It is flanked on the east by the Rockall Trough and to the west by the abyssal plain of the Iceland Basin (Fig. 9.1). The main topographic features are the opposed crescent-shaped *Hatton Bank* in the west and *Rockall Bank* in the east, separated by the deeper *Hatton-Rockall Basin*. Rockall Rock at the northeast end of Rockall Bank, is the only emergent part of Rockall Plateau. The Rock is only 28 m by 25 m in extent and 20 m high.

Rockall Bank has been studied in greater detail than the other parts of the Rockall Plateau, both in terms of surface features and of deeper structure. In detail the bank surface is irregular and composed of solid bedrock outcrops, sand banks, boulders, gravel and cold water reefs. Because of its isolated position and the surrounding deep water, Rockall Bank is essentially free from present day terrigenous sedimentation and this has resulted in the accumulation of carbonate sediments.[1] These form a thin veneer on the varied coarse glacial material and rock outcrop. The shallow carbonate facies (to 120 m) comprise sand and gravel of rounded bryozoan fragments with serpulids. These sediments pass outwards (120–220 m) through a zone of sand grade mollusc and echinoderm fragments with benthic foraminifera, to an outer zone (below 220 m) dominated by pelagic foraminifera with one metre high living colonies of branched corals.

Hatton Bank is a physiographic feature lying along the northeastern margin of Rockall Plateau. It has an overall northeast-southwest trend which changes to east-west at the northern end. Much of the bank lies between 700 m and 1,200 m below sea level. A somewhat separate oval prominence on the southern extension of Hatton Bank at the western point of Rockall Plateau is called Edoras Bank. Between Rockall and Hatton Banks is an intervening trough of deeper water going down to 1,700 m: the *Hatton-Rockall Basin*. The basin is closed at its northern end by the circular prominence of the *George Bligh Bank* but plunges quickly at its southern end into the deep water of the North Atlantic.

Southeast of the shallower water complex of Rockall Plateau is the southwest trending *Rockall Trough*. The trough deepens from around 2,000 m in the neighbourhood of Anton Dohrn Seamount in the northeast to 4,000 m in the southwest whereafter the floor of the

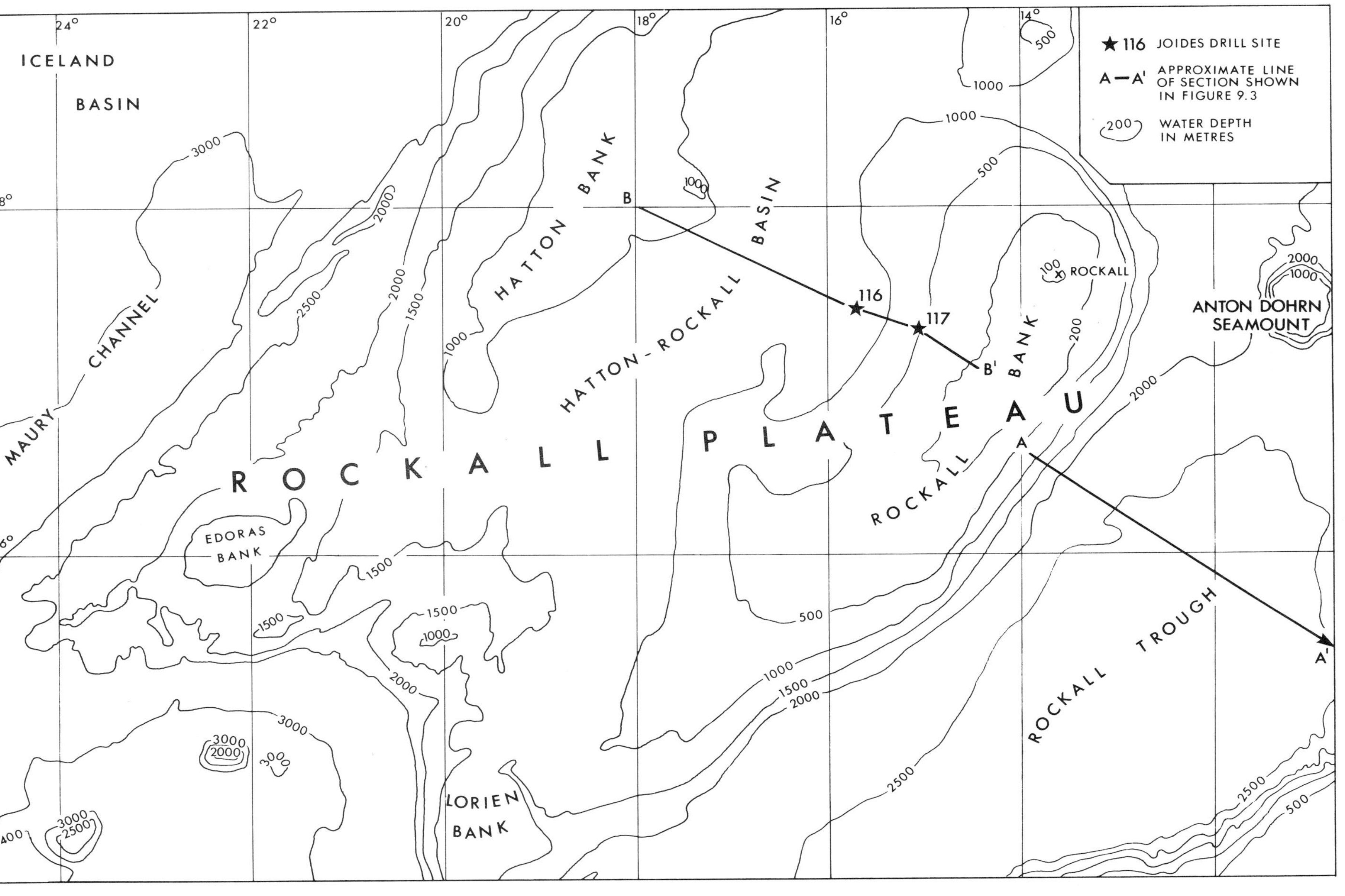

Figure 9.1. Morphology of the Rockall Plateau (taken mainly from GEBCO Map 5.04, 1978), showing the approximate lines of the cross-sections on Fig. 9.3).

trough descends quickly onto the Porcupine Abyssal Plain.

The physiography of the Rockall region reflects the underlying geology. Deep sedimentary troughs of Mesozoic-Tertiary strata underlie the deeper water areas but the younger strata are thin or absent over the intervening banks.

The critical position occupied by the Rockall region in any reconstruction of the history of the North Atlantic has been outlined in Chapter 1. The general juxtapositions were probably close to those shown on Fig. 1.3 although attempts at too precise a 'fit' at this stage of our knowledge are possibly unwise.

Geophysical Data

In the early 1970s seismic refraction and reflection programmes established velocities compatible with that of continental crust. The depth of the Mohorovicic discontinuity is about 30 km under Rockall Bank and 22 km beneath Hatton-Rockall Basin.[2–4] Only a thin veneer of sediment was found on Rockall and Hatton Banks but the reflection profiles revealed that Rockall Trough and the Hatton-Rockall Basin contain a thick sequence of layered sediments, displaying prominent reflectors, of Mesozoic and Tertiary age.[2–4]

Seismic reflection profiling has been carried out over the past decade and the lines form a network across Rockall Trough and Rockall Plateau, with particular emphasis on the Trough and Rockall Bank. Much of the early work, carried out by various academic research vessels,[3] was of moderate resolution. More recently[5] the results of multifold reflection lines in southern Rockall Trough have been published. The reflection seismic programmes have established a basic seismic stratigraphy above acoustic basement. Four seismic units are recognized bounded by, in ascending order above basement, reflectors Z, Y, X and R_4. Reflector R_4 is particularly well developed (Fig. 9.3) and is recognized from the Bay of Biscay to Rockall Trough.

In addition to the seismic work carried out for research purposes there have been a number of speculative[6] and proprietary programmes shot by the oil industry. However, the density of grid is much less than in most of the other basins described in this book. Hydrocarbon exploration has barely begun in this remote and difficult region.

Magnetic data[5, 7, 8] confirm three distinct provinces: firstly the Rockall Plateau itself of very varied magnetic character, secondly the oceanic area of the North Atlantic to the west associated with correlatable magnetic striping, and thirdly the undisturbed field of the sediment-filled Rockall Trough.

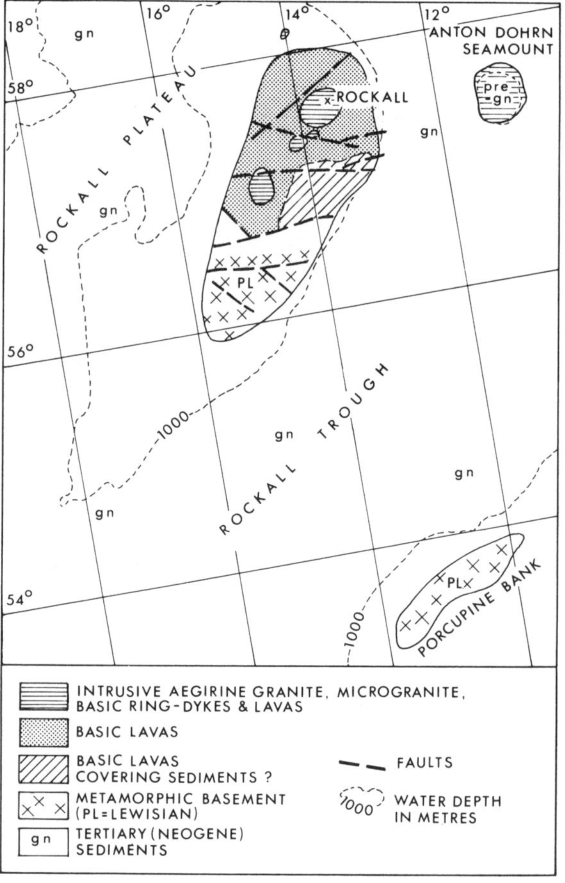

Figure 9.2. Sub-Pleistocene geology of the Rockall Plateau and Rockall Trough (in the main after I.G.S. 1:2.5 million scale map, and Roberts and Jones (1978)[8]).

Stratigraphy

Hatton-Rockall Basin

Up to 3,000 m of near-horizontal sediments infill the Rockall-Hatton Basin. The age of this sedimentary infill is partly established through correlation with the Joides 116 and 117 boreholes (figs. 9.1 and 9.4). Seismic profiles in the basin show 520–715 m of relatively unconsolidated sediments with velocities from 1,700 to 2,400 m/s which correspond to the Middle-Upper Tertiary oozes and chalks of the Joides site 116 (above the R_4 reflector). They are underlain by a layer some 1,220–1,525 m thick, with a velocity of 2,880 m/s which in its upper part corresponds to the Eocene-Palaeocene of the Joides 116 and 117 sites. The underlying portion of the layer may represent sediments or volcanic lavas which rest on a high velocity (6,700 m/s) layer than can be correlated with the crust beneath Rockall Bank.

The Joides bore holes (Fig. 9.4) provide virtually the only deep drilling data available in the area.[9] Site 116 was drilled in 1,174 m of water in the centre of Hatton-Rockall Basin and penetrated some 808 m of Oligocene and Mio-Pliocene ooze and chalk with stringers of chert, and 46 m of Upper Eocene carbonate ooze. Site 117 was drilled in 1,052 m of water on the western flank of Rockall Bank and penetrated some 152 m of Oligocene cherty limestones and ooze overlying 152 m of Lower Eocene ooze, sandstone and clay. Seismic basement of plateau basalt was encountered at the base of the section and the hole reached a total depth of 353 m below the sea bed.

Within the centre of the basin the Upper Tertiary beds overlie Oligocene and Lower Miocene cherty limestones, which themselves rest unconformably on a relatively thin sequence of well consolidated pre-Eocene sediments. The age and nature of the sediments underlying the Eocene unconformity throughout most of the basin is somewhat conjectural. The only indication of their character is found in the section of Palaeocene graded clays and sands sampled at the base of the drillhole Site 117, close to the edge of the basin. Because of the marginal situation of this sample, it may not be representative of the majority of the section which infills the centre of the basin. However, the tectonic history of the depres-

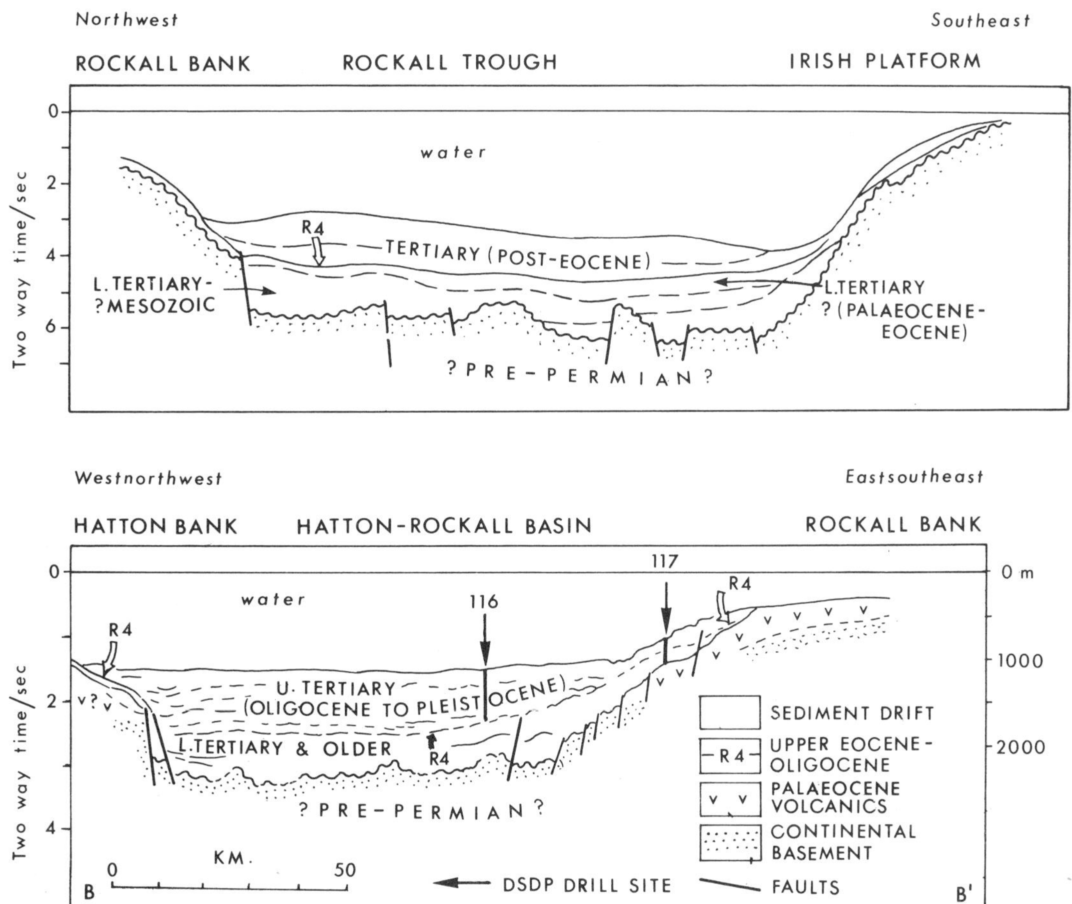

Figure 9.3. Sections based on reflection seismic profiles across: A-A₁ Rockall Trough (modified after Naylor and Mounteney (1975)[23]). B-B₁ Hatton-Rockall Basin (with modification after Roberts (1975)[14]). See Fig. 9.1 for position of section lines.

sion, with an early rifting phase, suggests the possibility that much of the basal sequence is composed of coarse continentally-derived sediments such as sandstones and conglomerates, laid down under shallow water or terrestrial conditions. The crust beneath Hatton-Rockall Basin is probably slightly thinner than beneath the adjoining banks, but it remains continental in thickness and character.

Rockall Trough

Seismic reflection profiles between Ireland and Rockall Plateau show that the present day topographic basin of Rockall Trough is only a relic of an originally much deeper structural feature now partially infilled by more than 3,000 m of roughly horizontally bedded strata. The greatest thickness of sediments within the basin occurs on the eastern side, close to the edge of the Irish-Scottish continental platform, where continuous deposition is thought to have taken place since the trough was first created.

Seismic reflection data[3] across the basin show that the infilling sequence of Tertiary and Mesozoic sediments contains three prominent reflecting horizons, R_4, X and Y with the oldest sediments comprising the pre-Y interval. This latter interval is further subdivided within the trough by reflector Z which separates an upper acoustically transparent unit (? pre-Late Palaeocene volcanics and sediments) from a lower unit with strong reflectors. This lower unit rests on basement, shows internal prograding towards the axis of the trough and may be a product of the early rifting phase.

Regional evidence suggests that Reflector Y corresponds to a Maastrichtian to Late Palaeocene age hiatus. In Rockall Trough the reflector is immediately below widespread lava flows (interpreted from seismic records) thought to be of Upper Palaeocene age.

Thus although the age of the basal sediments is uncertain, it seems likely that they will date back from Upper Cretaceous times to the time of formation of the trough as an embryonic rifted structure in the Permo-Triassic. Such sediments will have been laid down in terrestrial or shallow marine environments and will be characterized by irregular units of fluviatile and lacustrine sandstones, conglomerates and mudstones. The continental deposits will probably pass up into a layered sequence of shallow water carbonates and evaporites.

After the first phase of widespread crustal subsidence in the Northeast Atlantic, the floor of Rockall Trough (in particular the southern part), sank and a regime of moderately deep water sedimentation was established. Elsewhere over southern Britain and the Southern North Sea, the same subsidence is represented by a major break in Cretaceous sedimentation, and marks the sudden appearance of the extensive chalk seas. From the geophysical character and apparent thickness of the sedimentary interval between the lower and middle reflectors (X and Y), it seems likely that this marine environment extended westwards across parts of Ireland and the Celtic Sea to deposit a similar Upper Cretaceous chalk-greensand-clay sequence in the Rockall Trough.

The intermediate unconformity (Reflector X) in Rockall Trough is believed to correspond to a mid-Eocene period of widespread subsidence and decreased rate of sedimentation, and appears as a fairly strong continuous reflector. In the axial portion of Rockall Trough the X Reflector is in apparent conformable sequence with Y below and R_4 above.

The Palaeocene age of the Y reflector mentioned above and the well-established Eocene to Oligocene age for the R_4 reflector place constraints on the possible age of the X horizon. Towards the margins of the basin, however, X pinches out against Y and is itself cut by R_4. Thus in the marginal zones there may have been a prolonged hiatus in the Tertiary between Palaeocene and Miocene times.

Consequently the thickest and most complete sequence of post-Upper Eocene deposits is developed only within the centre of the basin, while towards the margins younger sediments overlie the unconformity. Lithologically the sediments are likely to be semi-consolidated limestones and calcareous oozes comparable to those encountered in drilling the Hatton-Rockall Basin, although they may contain a higher proportion of coarser land-derived material such as sand and silt along the eastern margin of the trough.

The uppermost unconformity (R_4) indicates a second period of Tertiary subsidence within Rockall Trough related to a change in the pattern of sea-floor spreading in the North Atlantic during the Oligocene. A similarity in the seismic characteristics of the reflectors overlying this unconformity with those identified in the Hatton-Rockall Basin, suggests that the sequence may also be composed of Oligocene cherts overlain by Miocene to Recent sediments. Towards the basin margins the cherts pinch out against the R_4 reflector and are overlapped by younger sediments. The post-Oli-

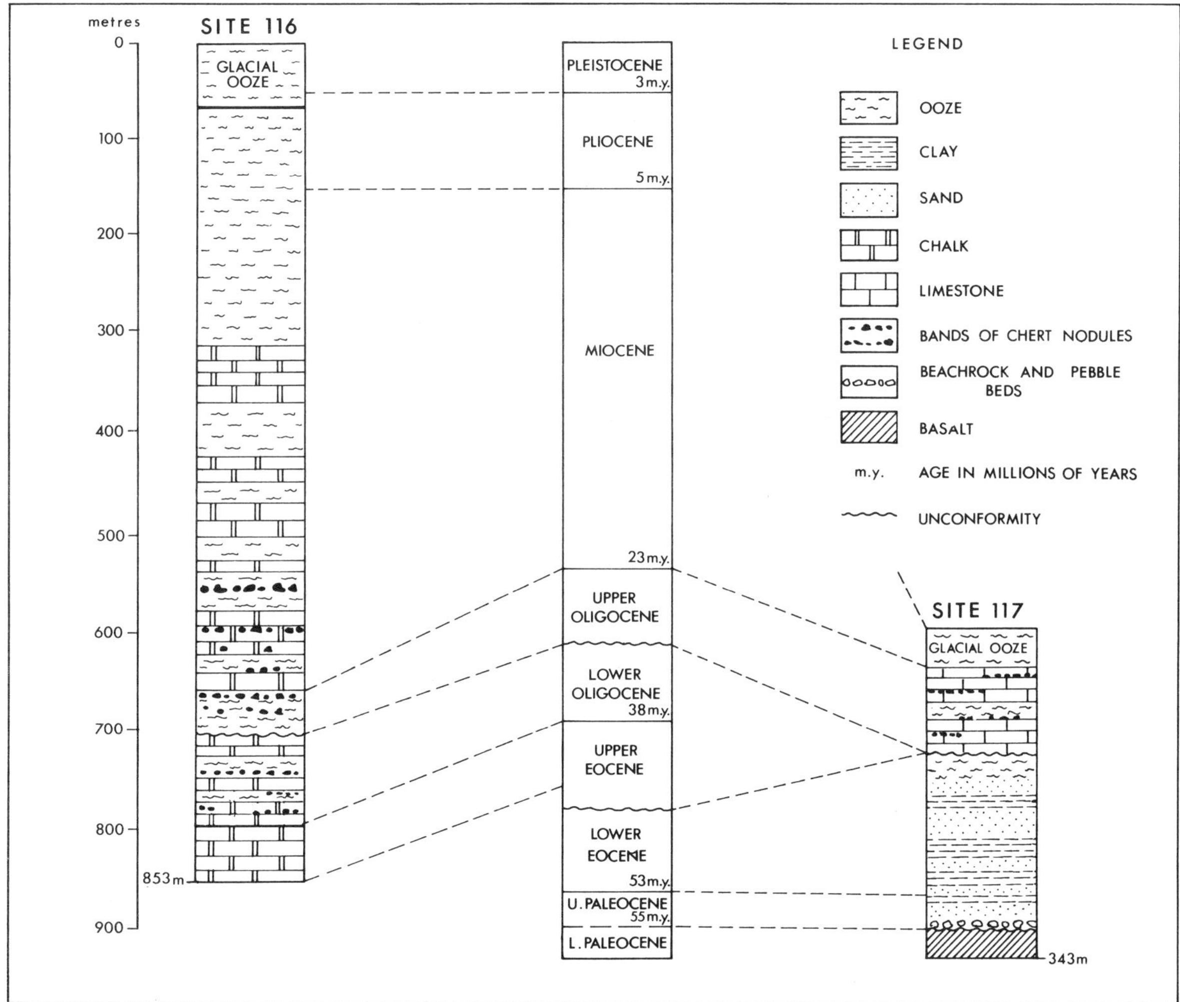

Figure 9.4. Stratigraphic sections penetrated by two deep water Joides boreholes on Rockall Plateau (for borehole locations see Fig. 9.1 and Fig. 9.3 Section B – B_1).

gocene stratigraphy is characterized by quieter deposition on the west in contrast to the fan deposition predominating along the eastern margin.

Rockall Bank

Rockall Bank has a thin cover of sediments resting on acoustic basement. Dredging and drilling on the bank south of 56°40′N has yielded Laxfordian (about 1,600 My) and Grenvillian (about 1,000 My) granulites.[10–12] Many regional reconstructions show the eastward extension at the Grenville Front (Fig. 1.4) as crossing the southern part of Rockall Bank along an east–northeast-trending magnetic lineament.[13–16] Rockall itself[17] is the aegirine granite (52 ± 9 My) tip of a large intrusive complex composed of Tertiary acid and basic igneous rocks. The magnetic picture for the bank is further complicated by the presence of Cretaceous microgabbros and by widespread basaltic extrusions such as that penetrated in the Joides 117 drillhole (Fig. 9.3). It is probable that much of the northern part of the bank (north of 56°40′) beneath the thin sediment cover is covered by basic lava flows whilst the southern portion is underlain by the Precambrian granulites (Fig. 9.2).[8]

Summary

Both Hatton and Rockall Banks have a shallow acoustic basement of continental crust. Rockall Bank is covered by only a thin veneer of Upper

Tertiary and Recent sediments while Hatton Bank is overlain by a slightly thicker Upper Tertiary sequence. Miocene sediments were deposited across the entire surface of Rockall Plateau following the phase of Upper Oligocene-Lower Miocene subsidence. Rockall Trough has a thick sedimentary infill and a more continuous history of sedimentation than the Plateau. Originally it may have been a down-faulted continental graben structure possibly as old as Permo-Triassic. If this is the case then continental type sediments may be present at depth, particularly in the southern part of the Trough. The introduction of a marine environment into the Rockall Trough during the Late Mesozoic may have resulted in the deposition of marine sands, shales and limestones. At the same time crustal extension was possibly taking place along the line of Hatton–Rockall Basin. This led to a crustal sag, but although it appears that the continental crust has been thinned there is no indication of oceanic crustal material. As we have seen, seismic profiles and the Joides drilling results confirm that Hatton-Rockall Basin also contains a thick sequence of Mesozoic and Tertiary rocks.

The rapid northward extension of the mid-Atlantic Ridge between the North American plate and the Greenland, Rockall Plateau and European plate led to considerable subsidence over the adjoining continental shelves. Deposition of Upper Cretaceous marine sediments in the Rockall Trough, resting on shallow water dolomitic calcarenites, carbonate sands and shelly limestones, probably resulted from the subsidence. This can be inferred from study of the Joides borehole 111 on Orphan Knoll (Newfoundland: see Chapter 10) which prior to the opening of the Labrador Sea was adjacent to Rockall Plateau (Fig. 1.3).

The break-up of the Rockall-Greenland plate was accompanied by the intrusion of Lower Palaeocene granites (including Rockall Rock itself) and the extrusion of basalts which were subaerially weathered. This volcanic activity persisted through to the Lower Eocene. Ocean floor spreading about the Reykjanes Ridge began to slow down, producing a period of widespread subsidence in the Upper Eocene and Lower Oligocene which was accompanied by the deposition of the marine clays and oozes seen in the Joides 116 and 117 boreholes. Deep water conditions then became prevalent across the Hatton-Rockall Basin and Rockall Trough and resulted in the deposition of the cherts and oozes of the Oligocene and Miocene. During the Tertiary, sedimentation spread from the centre of the basins onto the adjoining Rockall and Hatton Banks.

During the early phase of the opening of the North Atlantic crustal distension occurred along the line of Rockall Trough. Permo-Triassic rifting was probably followed by a major phase of rifting in Late Jurassic-Early Cretaceous time. It has been considered by a number of authors[15, 16, 18–20] that this period of crustal thinning was followed by the development of an axis of spreading along the trough which produced new ocean floor and separated a Rockall 'microcontinent' wholly or partly from the European Plate. Several models of North Atlantic reconstruction have incorporated this concept. However, more recent work has thrown doubt on the existence of oceanic crust along the length of the trough; the evidence points to much crustal thinning but no indication of oceanic crustal material. The main axis of Atlantic spreading did not penetrate northward to the Rockall region until post-Albian time. The crust west of Goban Spur is post-Albian in age as evidenced by the fact that known Albian and older sediments are shown on seismic data not to continue onto the crust in this region.[5] The spreading axis became extinct in this area at 75 My and moved westwards to the Labrador Sea.

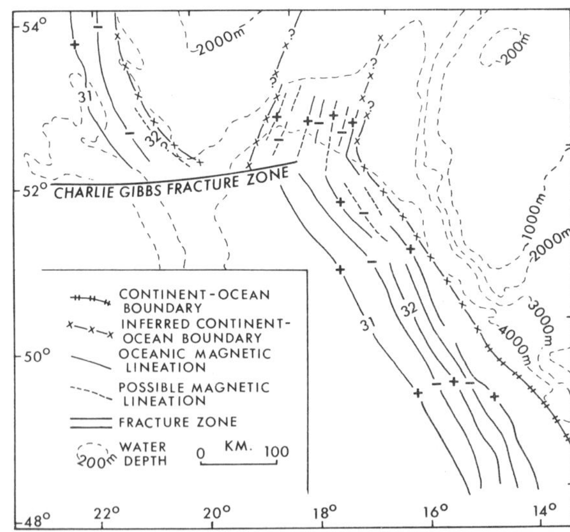

Figure 9.5. The location of oceanic magnetic lineations (31 and 32) at the mouth of Rockall Trough (selected data after Roberts, Masson and Miles (1981)[5]).

The position of anomalies 31 and 32 across the mouth of Rockall Trough is shown on Fig.

9.5. Anomaly 32 is the oldest clearly defined magnetic strip in this area (Fig. 9.3). According to some authorities[5] this is 60 km west of the contact between oceanic and continental crust based on geophysical data. Weaker discontinuous linear magnetic anomalies in the intervening zone have been tentatively identified as anomalies 33 and 34. If this is accepted, then these anomalies can be considered to pass east of the end of the Gibbs Fracture Zone to link with weak axial north-south anomalies in the mouth of Rockall Trough. This would imply a pre-32 (75 My) anomaly spreading history for at least the southern part of Rockall Trough. However it should be noted that the supposed 33 and 34 anomalies are weak and discontinuous and occur along the oceanic-continental interface. This can be expected to be a complex zone of thinned and broken continental crust with possible linear intrusive bodies inherited from the pre-rifting phase and likely to complicate the magnetic picture. The reader is warned of a further complication in the literature in that some authorities interpret the North Atlantic anomaly data differently, such that anomalies 31 and 32 of this account become anomalies 33 and 34.[18, 20] However, this reassignment does not affect the immediate discussion.

The weak magnetic anomalies at the mouth of Rockall Trough between 52°0′ and 52°30′N are separated by a complex suite of west northwest-trending anomalies from the magnetic quiet picture which characterizes the main part of the trough to the north. One suggestion is that the quiet zone may represent the thicker and possibly changed nature of the oceanic crust in this latter area, as postulated also for the region north of the Anton Dohrn Seamount (Fig. 9.1).[21] This oceanic crust is supposed to have been formed during the Cretaceous normal polarity interval and therefore did not develop distinctive magnetic stripping.

As mentioned above, it has been suggested that ocean floor spreading occurred along the Rockall Trough and also further to the northeast along the Faeroe-Shetland Channel. However there is now considerable evidence that Mesozoic rocks are present beneath the eastern margin of the Faeroe-Shetland Channel and some workers[22] prefer the concept that the Channel is floored by this foundered continental crust, possibly with linear igneous activity along the axis. A similar picture could certainly hold for the main part of the Rockall Trough.

It has been argued that the reconstructions of the North Atlantic by various workers incorporate some degree of oceanic floor spreading beneath Rockall Trough and therefore give independent collaboration to this view. However, any review of the recent literature on this subject shows a wide range of opinion regarding the details of the reconstruction. Many of the reconstructions contain unresolved problems. For example the reconstruction reproduced in Fig. 1.3 of this book (which involves some spreading in Rockall Trough) results in some continental overlap and its author[19] comments that 'The palaeogeographic position of the continents at the time of their separation as given here, shows considerable overlap between their boundaries. The position of these boundaries is not known with precision everywhere and an accurate description of these boundaries may reduce some of the apparent misfits.' It is also true to say that crustal stretching is a difficult factor to incorporate in North Atlantic reconstructions.

At the present time, therefore, the geophysical evidence for the existence of oceanic crust along Rockall Trough must be regarded as ambiguous and not capable of resolution. It seems likely, on balance, that extreme crustal extension, with related linear intrusions, occurred but that a full spreading axis was never established.

Hydrocarbon Potential

The question of the oil and gas potential of the Rockall Trough and the Hatton-Rockall Basin can only be addressed in very general terms. The results of only a small amount of high-resolution seismic profiling have been published, and drilling information is limited to the two Joides holes. It must be clear from the account given above that little is known concerning the pre-Tertiary stratigraphy of the basins.

Reservoir potential may exist in the Permo-Triassic continental deposits presumed to exist along the basins, and in Jurassic-Lower Cretaceous sandstones. The basin margins, with active fault movement and pregrading fan and delta sequences, probably afford the best opportunities for significant thicknesses of sandstone reservoirs. Migration of hydrocarbons from post-Triassic source beds may also have occurred towards the basin margins, where wedge-out and drape over fault blocks provides entrapment conditions. The problem of source bed maturity is likely to exercise significant control on the search for hydrocarbons in these basins. It is certainly true also that the fine-grained Neogene sequences are too shallow and lacking in reservoirs to offer significant exploration potential.

The first well was drilled in the north Rockall area in 1980. BNOC operated for a consortium of 19 companies plus the UK Department of Energy. The water depth was 1,355 m. At the time, this was the deepest water ever drilled in Europe and the second deepest in the world. No results have been released regarding the geology encountered in the well although scouting reports suggest that it encountered a thick igneous section.

References

1. SCOFFIN, T. P., ALEXANDERSSON, E. T., BOWES, G. E., CLOKIE, J. J., FARROW; G. E. and MILLIMAN, J. D. 1980. Recent Temperate sub-photic carbonate sedimentation Rockall Bank, Northeast Atlantic. *J. Sed. Petrol.* **50**, 331–356.
2. SCRUTTON, R. A. and ROBERTS, D. G. 1971. Structure of Rockall Plateau and Trough, northeast Atlantic: *in* Delany (*Ed.*). *The Geology of the East Atlantic Continental Margin.* Rep. Inst. *Geol. Sci.* 70/14, 77–87.
3. ROBERTS, D. G. 1975. Marine geology of the Rockall Plateau and Trough. *Philos. Trans. R. Soc. London* **A 278**, 447–509.
4. SCRUTTON, R. A. 1972. The crustal structure of the Rockall Plateau microcontinent. *Geophys. J. R. astron. Soc.* **17**, 259–275.
5. ROBERTS, D. G., MASSON, D. G. and MILES, P. R. 1981. Age and Structure of the southern Rockall Trough, new evidence. *Earth planet. Sci. Lett.* **52**, 115–128.
6. NAYLOR, D. 1972. *The hydrocarbon potential of Western Britain and Ireland.* North Sea Conferences 1 & 2 IPC Industrial Press, 130–137.
7. ROBERTS, D. G. 1970. Recent geophysical studies on the Rockall Plateau and adjacent areas. *Proc. Geol. Soc. London* **1662**, 87–93.
8. ROBERTS, D. G. and JONES, M. T. 1978. A bathymetric magnetic and gravity survey of the Rockall Bank, H.M.S. Hecla 1969. *Admiralty Marine Science publ. No. 19*, 45 pp.
9. ROBERTS, D. G., MONTADERT L. and SEARLE, R. C. 1979. The western Rockall Plateau—stratigraphy and structural evolution: *in* Initial reports of the deep sea Drilling Project 48. *U.S. Govt. Printing Office, Washington D.C.*, 1061–1088.
10. ROBERTS, D. G., MATTHEWS, D. H. and EDEN, R. A. 1972. Metamorphic rocks from the southern end of the Rockall Bank. *J. Geol. Soc. London* **128**, 501–506.
11. ROBERTS, D. G., ARDUS, D. A. and DEARNLEY, R. 1973. Precambrian rocks drilled from the Rockall Bank, *Nature Phys. Sci.* **244**, 21–23.
12. MILLER, J. A., ROBERTS, D. G. and MATTHEWS, D. H. 1973. Rocks of Grenville Age from Rockall Bank. *Nature Phys. Sci.* **246**, 61.
13. BAILEY, R. J. 1979. The Continental Margin from 50°N to 57°N; its geology and development: *in* Banner, F. T., Collins, M. B. and Massie, K. S. (*Eds*). *The North-West European Shelf Seas Vol 1.* Elsevier, Amsterdam, 11–24.
14. ROBERTS, D. G. 1975. Tectonic and stratigraphic evolution of the Rockall Plateau and Trough: *in* Woodland, A. W. (*Ed*) *Petroleum Geology and the Continental Shelf of North West Europe 1. Geology.* Applied Science Publishers, London, 131–148.
15. ROBERTS, D. G. 1975. *The solid geology of the Rockall Plateau: in* Expedition to Rockall 1971–72. *Rep. Inst. Geol. Sci. London* 75/1, 3–10.
16. ROBERTS, D. G. 1974. Structural development of the British Isles, the Continental Margin, and the Rockall Plateau: *in* Burke, C. A. and Drake, C. L. (*Eds*). *The Geology of Continental Margins.* Springer-Verlag, New York, 343–359.
17. SABINE, P. A. 1965. The geology of Rockall, North Atlantic *Bull. geol. Surv. Gt. Br.* **16**, 156–178.
18. KRISTOFFERSON, Y. 1978. Seafloor spreading and the early opening of the North Atlantic. *Earth Planet Sci. Lett.* **38**, 273–290.
19. SRIVASTAVA, S. P. 1978. Evolution of the Labrador Sea and its bearing on the early evolution of the North Atlantic. *Geophys. J. R. Astron. Soc.* **52**, 313–357.
20. TALWANI, M. 1978. Distribution of basement under the eastern North Atlantic Ocean and the Norwegian Sea: *in* Bowes, D. R. and Leake, B. E. (*Eds*) Crustal evolution in northwestern Britain and adjacent regions. *Geological Journal Special Issue No. 10.*
21. BOTT, M. H. P., ARMOUR, A. R., HIMSWORTH, E. M., MURPHY, T. and WYLIE, C. 1979. An explosion seismology investigation of the continental margin west of the Hebrides, Scotland at 58°N. *Tectonophyscis* **59**, 217–232.
22. RIDD, M. F. 1981. Petroleum geology west of the Shetlands: *in* Illing, L. V. and Hobson, G. D. (*Eds*). *Petroleum Geology of the Continental Shelf of Northwest Europe*, Heyden & Son Ltd, London, 414–425.
23. NAYLOR, D. and MOUNTENEY, S. N. 1975. *Geology of the North-West European Continental Shelf Vol. 1.* Graham, Trotman & Dudley London, 162 pp.

Chapter 10

Eastern Canada

Introduction

The opening of the North Atlantic, described in the first chapter, carries with it the implication that the continental shelves of the opposing coastlines had a comparable geological history during a substantial portion of the Mesozoic Era. Published data resulting from oil exploration on the eastern Canadian Shelf represents the most pertinent stratigraphic control for the relatively unexplored extreme western seaboard of Europe. Prior to opening of the Atlantic these offshore points were closer together than were the westernmost basins of Europe, such as the Porcupine Basin, to the onshore sections exposed in southern England or northern France. In this chapter therefore we attempt to outline the geology of the eastern Canadian continental margin for the purposes of comparison with the basins of Ireland and West Britain. This is not intended to be an exhaustive account and the reader is directed to the substantial literature, part of which is listed at the end of this chapter.

Much of our current knowledge of the geology of the East Canadian continental margin has been acquired during the last fifteen years, in the main through petroleum company exploration activity. Valuable geophysical surveys – magnetic, gravity and seismic – have also been conducted by a number of academic institutions. This was particularly the case in the early phases of shelf exploration and also in the continuing effort to understand the deeper structure of the continental margin and pre-drift fits and correlations around the North Atlantic. Several holes drilled as part of the Joides Deep Sea Drilling Project helped to clarify the structure and history of the Newfoundland-Labrador section of the shelf. Since the late 1960s approximately 150 wells have been drilled in the East Canadian offshore area. This substantial drilling programme at first achieved little success. However, in 1979 the Hibernia Oilfield was discovered offshore from Newfoundland and since that time, with quickening drilling activity, a number of further discoveries have been made.

The result of the exploration activity has been the recognition of four major Mesozoic-Cenozoic basinal provinces off eastern Canada[20] south of 60° North latitude (the Baffin Island and Greenland basins are not considered here). From north to south these are: the *Labrador Basin*, the *East Newfoundland Basin*, the *Grand Banks Province* and the *Scotia Basin* – as shown on Fig. 10.1. A number of named positive elements separate the main basins and a variety of names have been given in the literature to sub-basins (or basins) within this Margin complex.

The East Canadian shelf is commonly 200–300 km broad. Considerable areas of the shelf are in the 200–500 m water depth range, and at its outer margin there is rapid deepening to 2,000 m. Operating conditions for exploration are harsh and drilling costs are high in consequence. An additional problem on the Labrador and Newfoundland shelves are the icebergs

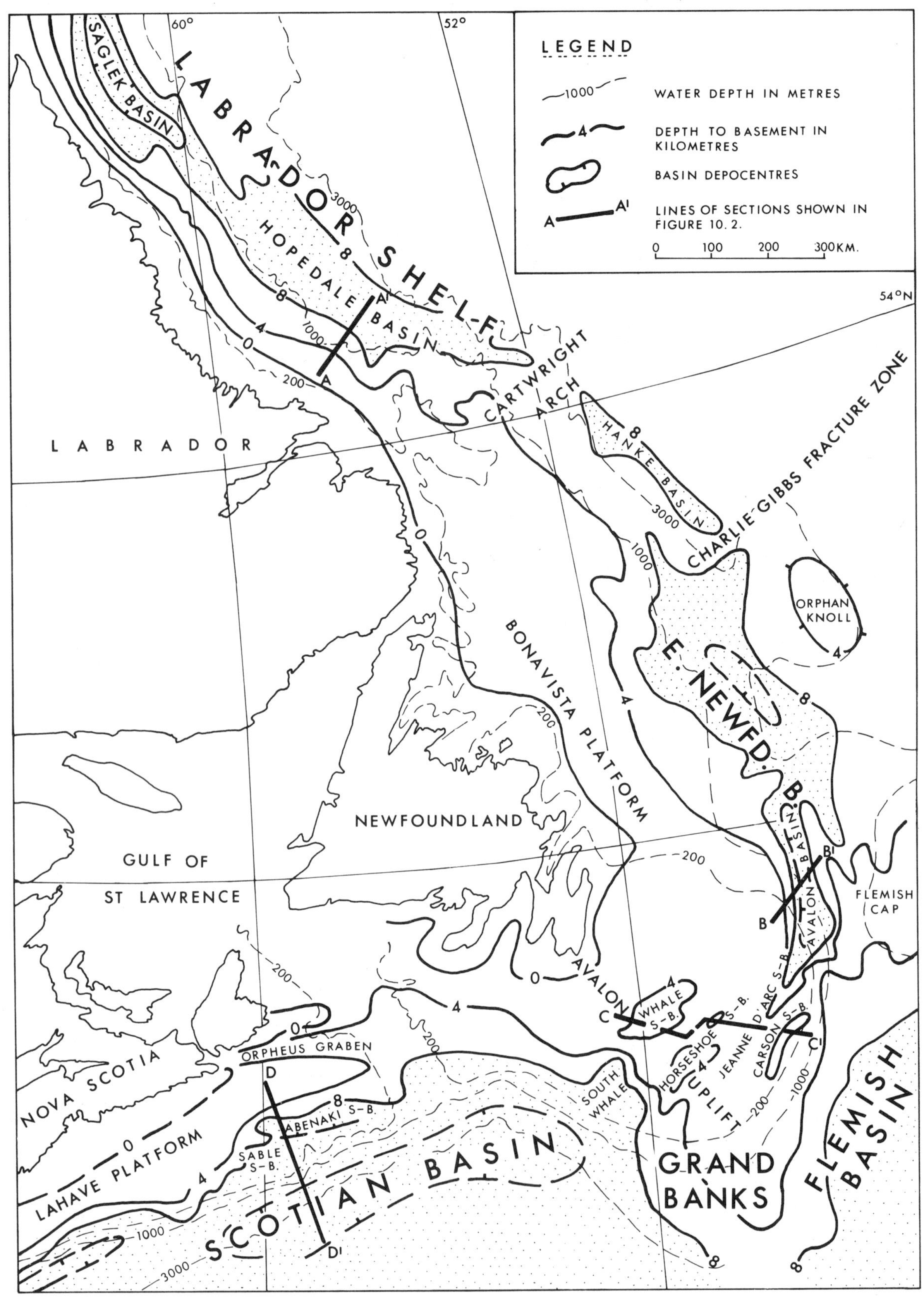

Figure 10.1. Map showing the sedimentary basins offshore eastern Canada (in the main after Geological Survey of Canada Map 1900A 1977). Sedimentary thicknesses in the Gulf of St. Lawrence not shown. Lines of section A – A′ to D – D′ shown in Fig. 10.2.

drifting south from Greenland. Whilst some of these may be diverted by the use of tugs others have caused rigs to move hurriedly off location.

Stratigraphy

Pre-Mesozoic Rocks and Structure

The maritime area of eastern Canada, including the continental margin, has a wide variety of pre-Mesozoic rocks at outcrop or in the subsurface. The region has a complex Precambrian and Palaeozoic history and it is not within the scope of this book to enter into a detailed description of this. However, in order to set the scene for our analysis of the Mesozoic basins of the area a brief resumé of the main aspects of the pre-Mesozoic evolution is given. The most significant geological feature of the area is the North Appalachian chain which extends through the Maritimes to Newfoundland. The inception of the Appalachian zone as a geosynclinal trough in the Late Precambrian was followed by folding in the Early Palaeozoic. Flanking troughs developed to the northwest (Acadian) and southeast (Meguma) of the folded stable Avalon Platform. Folding and intrusion of the Acadian geosyncline then followed during the Ordovician (Taconic) orogeny. The whole area was then uplifted as a stable basement platform.

Upper Palaeozoic sedimentation (Late Devonian to Pennsylvanian) was controlled by a series of northeast-trending faults which defined horst and graben structures. A basal coarse redbed sequence grades up into finer redbeds, shales and limestones with thick evaporite units. The Upper Carboniferous comprises a varied sequence of red and grey clastics with coal-bearing strata.

Many authors have discussed the problems of the fit of Europe and Canada prior to seafloor spreading. The reconstruction used on Fig. 1.3 in Chapter 1 seems to afford good correlation with known facts without attempting too tight or inflexible a fit of the margins. It has been demonstrated by magnetic studies that the structural zones of the Appalachians extend eastwards to the outer edge of the Canadian margin.[1, 2] The overlying Mesozoic sedimentary basins are separated by structural highs which imply rejuvenation of Precambrian and Palaeozoic structures.

Labrador Basin

The Labrador Shelf[3–7] is covered by a wedge of Mesozoic and Tertiary strata. Precambrian rocks of the Canadian Shield, which are at outcrop on the mainland, dip seawards beneath the shelf. Deep drilling has revealed that Ordovician and Carboniferous rocks are present over the southern part of the shelf, at least within the 1,000 m isobath, with Precambrian rocks occurring farther north. The overlying Mesozoic sequence, which is more than 10 km thick, has been divided into six formations (Fig. 10.2 and Table 10.1) with the oldest being Early Cretaceous (Berriasian or Valanginian to Barremian) lava flows and subordinate sediments of the *Alexis Formation*. Overlying continental deltaic sandstones and conglomerates, shales and coals range in age up to Albian (*Bjarni Formation*). They are succeeded unconformably by marine shales with basal sandstones (*Markland Formation*) of Late Cretaceous to Early Palaeocene age which are then overlain by claystones with minor turbiditic sandstones and carbonates (*Cartwright Formation*: Palaeocene to early Eocene). Three thick Eocene to Pliocene sandy and silty marine mudstone formations (*Kenamu, Mokami and Saglek Formations*) complete the section.

The Labrador Shelf is underlain by a graben and horst structure in the basement and this resulted in variation in age and thickness in the oldest Mesozoic rocks. The younger Mesozoic formations are more uniform in type and thickness, as might be expected by marine sedimentation gradually smoothing out the buried fault-controlled topography.

TABLE 10.1
Mesozoic and Tertiary formations in eastern Canadian offshore basins

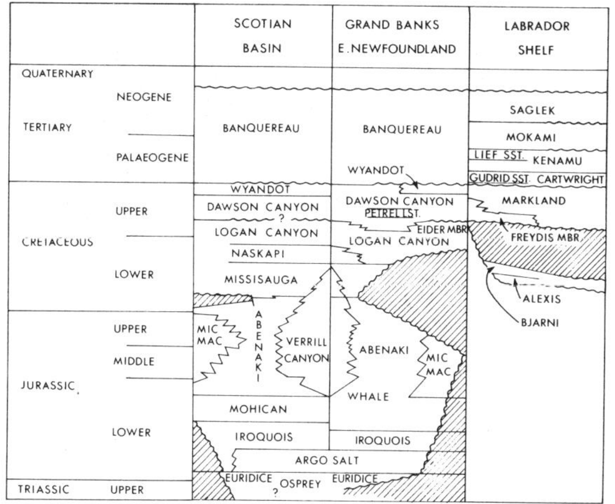

East Newfoundland Basin

The continental shelf immediately east of Newfoundland[8–10] comprises a platform area with

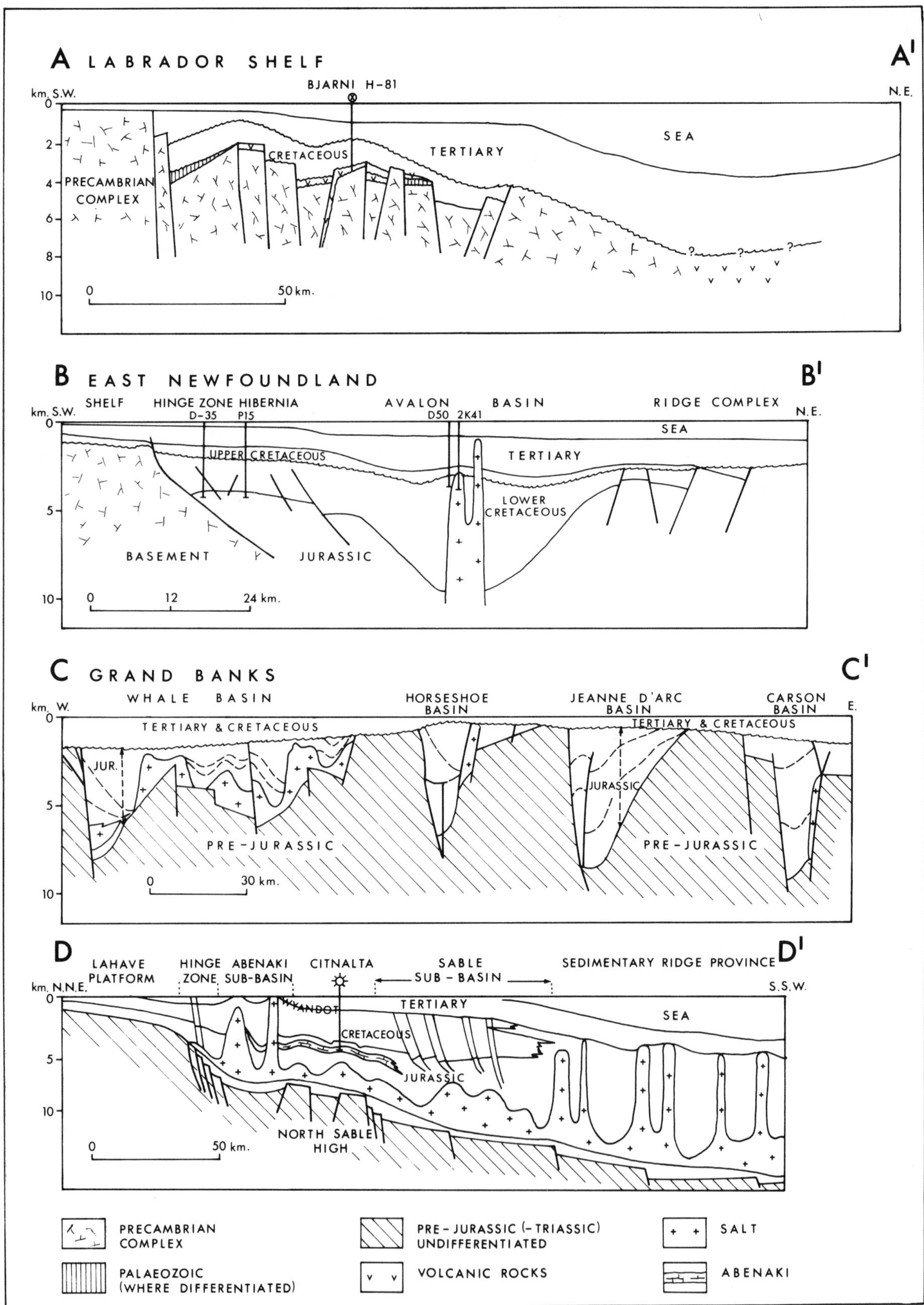

Figure 10.2. Diagrammatic cross-sections of the East Canadian sedimentary basins: see Fig. 10.1 for locations. Sections A – A′ (Labrador Shelf) and D – D′ (Scotian Basin) with modifications after Purcell *et al.* (1980)[6]. Section B – B′ (East Newfoundland) with slight modification after McKenzie (1981)[10], and C – C′ after Amoco Canada and Imperial Oil (1973).[14]

relatively thin (less than 2 km) Mesozoic cover – the *Bonavista Platform* (Fig. 10.4). The platform basement complex contains Precambrian to Palaeozoic rocks. On the eastern margin of the platform there is a hinge zone into the depositional thick (up to 12 km of sediments) area of the *East Newfoundland Basin.* The basin extends southwards as the *Avalon Basin* and narrows towards the *Jeanne d'Arc Graben* of the Grand Banks area, with an outer *Ridge Complex* on its eastern side (Fig. 10.1). The East Newfoundland Basin is bounded by the *Flemish Cap High* and *Avalon (Grand Banks) Uplift* in the south sector and by the ocean basin and the *Orphan Knoll – Flemish Cap* elements to the northeast. To the northwest the Charlie Gibbs Fracture Zone forms a distinct boundary, although the isopachytes for the Mesozoic cover link around the western end of the Fracture Zone with those of the Labrador Basin to the north.

Drilling on the southeastern margin of the basin has revealed redbeds of probably *Triassic* age. Progressive subsidence resulting from continental distension gave rise to marine incursions into the basin and deposition of evaporites including thick salt sequences. A further transgressive pulse in the *Early Jurassic* led to the deposition of dolomites and sabkha anhydrites of the *Iroquois Formation.* There was apparently connection across the Grand Banks area into the southern basins at this time. Deposition of *Middle Jurassic* shallow water carbonates, sandstones and shales followed. During the *Late Jurassic* open marine carbonate and shale deposition, with basinal shale equivalents, continued until terminated by a tectonic episode in latest Jurassic time. Uplift, activation of positive areas and regional emergence followed. The shore face and delta systems prograded eastwards into the basin at this time. The transition zone between the Jurassic and Lower Cretaceous, the *Jeanne d'Arc Zone,* contains fluvial and shoreline sandstones and conglomerates. Overlying this is a *Lower Cretaceous* section consisting of siltstones and shales with interbedded sandstone members and minor limestones. Towards the base is the important *Hibernia Sandstone* unit (Fig. 10.4B) and beneath the mid-Cretaceous unconformity (Albian-Aptian) the *Avalon Sandstones* are prime exploration targets in the basin and contain oil in the Hibernia discovery area (discussed below). Transgression in the *Late Cretaceous* marked a return to glauconitic sandstone, siltstone and shale deposition in open shelf conditions, with minor carbonate. The exception is the well-defined period of carbonate deposition represented by the *Petrel Limestone* unit.

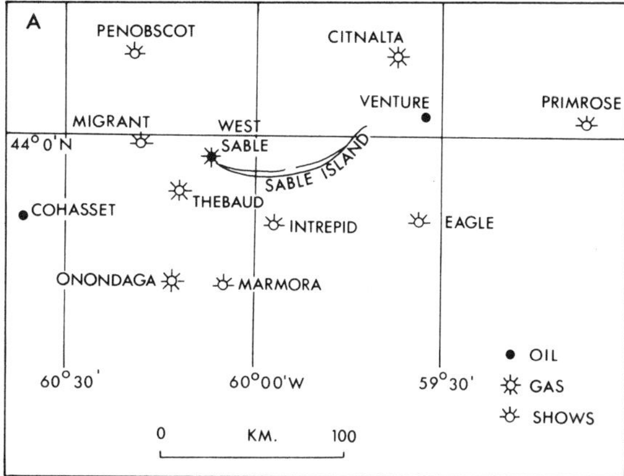

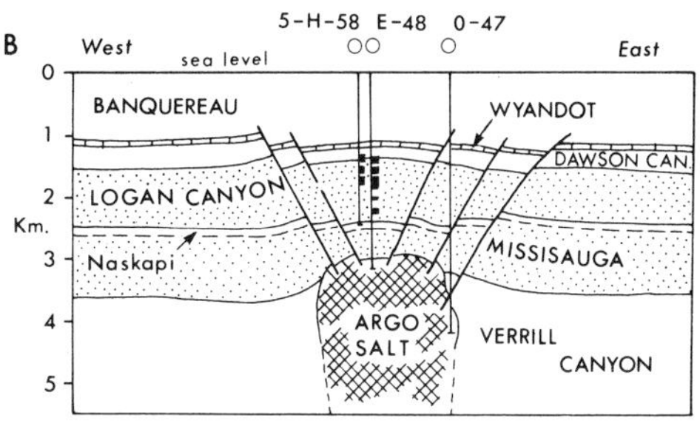

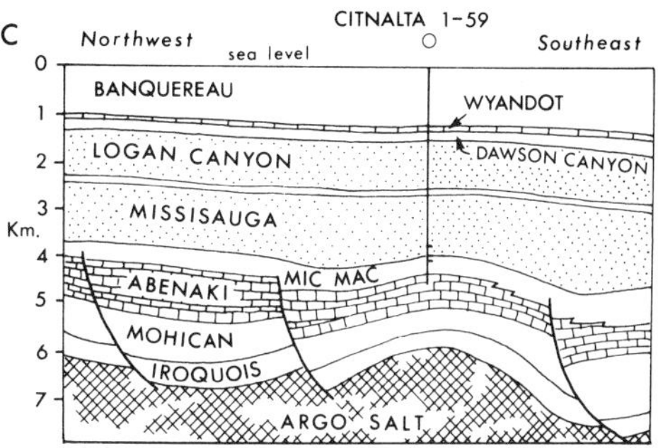

Figure 10.3. A. Location map of wells which have encountered oil and gas in the Sable Island area, Scotian Basin. Dry holes not shown.
B. Diagrammatic cross-section of the West Sable structure, showing some of the well control.
C. Diagrammatic cross-section of the Citnalta structure. Cross-sections with slight modification after Purcell *et al.* (1980)[6]. Black areas on the wells indicate generalized hydrocarbon zones.

Progressive pulses of marine advancement pushed the shoreline westwards towards the position of the present coast, with consequent interfingering of shoreline and marine facies. The *Tertiary* is represented by the *Banquereau*

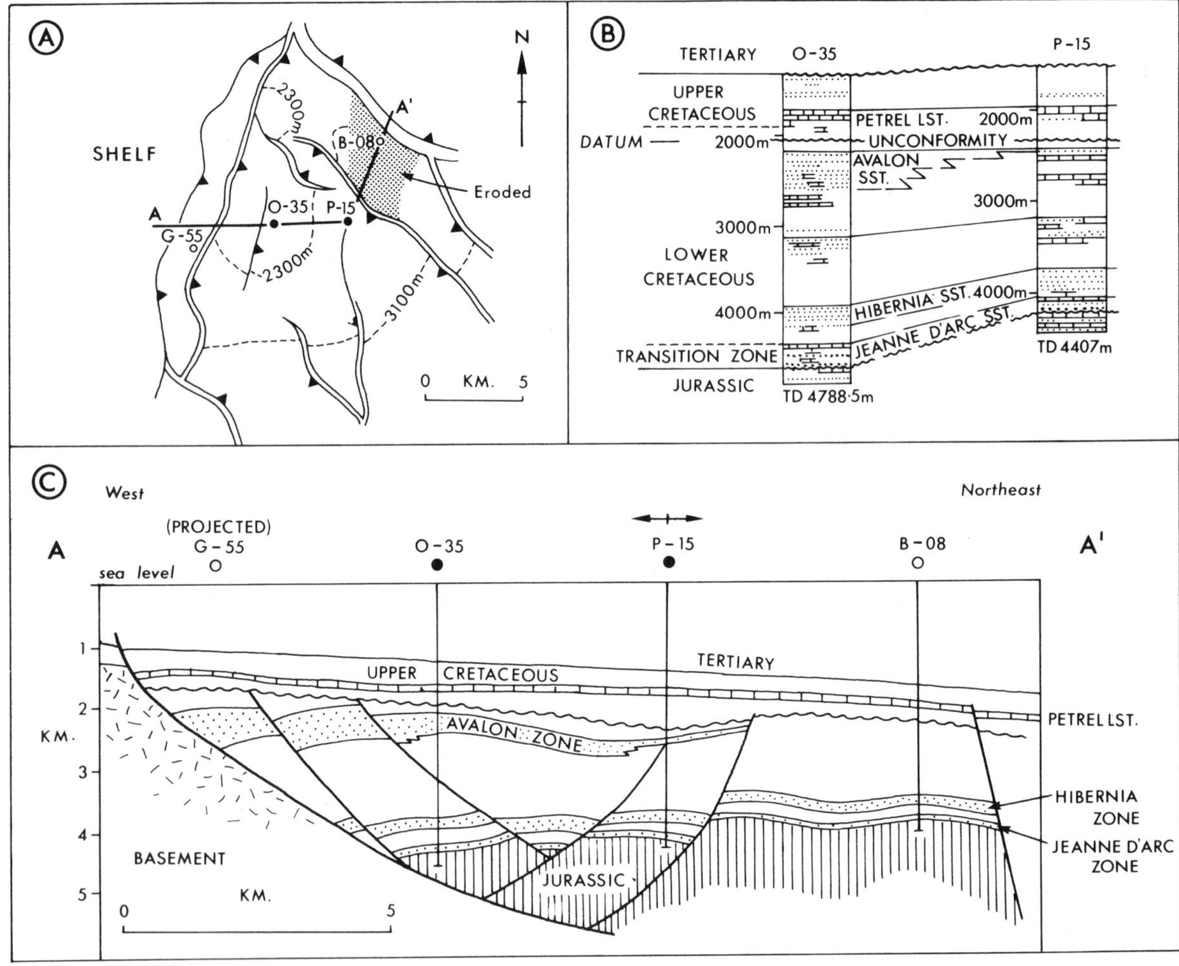

Figure 10.4. Hibernia Oilfield, East Newfoundland (Avalon) Basin.
A. Structure contours on the Lower Cretaceous Avalon Sandstone (for regional location see Fig. 10.2 B–B′).
B. Comparative Cretaceous stratigraphy of the O-35 and P-15 wells, using the mid-Cretaceous unconformity as datum.
C. Section A – A′, location shown on 10.4A above, across the Hibernia structure.
Figures taken with slight modification from McKenzie (1981).[10]

Formation, an extremely thick marine succession. In places the Tertiary attains a thickness of 8 km. The lower part of the formation comprises mudrocks, representing neritic conditions off the Newfoundland shelf and around Orphan Knoll, and bathyal conditions in the basin centre. There is a gradual coarsening upwards in the succession due to regression in the Late Tertiary.

Within the Avalon Basin salt diapirs rise from the Triassic-Lower Jurassic horizons up to varying levels in the section. Further east the little-known Ridge Complex overlies tilted basement fault blocks with intervening grabens. These latter contain a Lower Cretaceous-Jurassic section beneath the mid-Cretaceous unconformity.

The succession on *Orphan Knoll*, as evidenced by DSDP Site 111,[11] comprises 250 m of strata ranging from Jurassic to Plio-Pleistocene. Paralic carbonaceous grey sandstones and shales of Bajocian age are unconformably overlain by Albian and Lower Cenomanian (Cretaceous) glauconitic and sandy carbonates. Maastrichtian foraminiferal oozes follow upon a hiatus and are then overlain, after a similar break of sedimentation, by Eocene bathyal marls and clays and finally by Miocene-Pliocene and Pleistocene oozes and clays.

At the present time Orphan Knoll is an isolated topographic feature covered by 1,800 m of water and separated by a 3,000 m deep trough from the continental shelf. It lies 550 km northeast of the Newfoundland coast and affords interesting evidence relating to the opening of the North Atlantic. There is a history of deepening conditions with the marginal Jurassic sediments followed by Cretaceous shallow marine carbonates containing a fauna of European affinities. The European plate had clearly not separated far from Orphan Knoll at this time. There was rapid subsidence in the Maastrichtian and early Eocene and Orphan Knoll foundered rapidly. After this time the history was one of quiet deep water sedimentation.

Grand Banks Province

The Grand Banks Province[8, 12–14] comprises a series of fault-bounded northeast-trending sub-basins within the generally positive area off southeast Newfoundland, between the East Newfoundland Basin and the Scotian Shelf (Fig. 10.4). This north–northwest- trending positive element is usually termed the *Avalon Uplift*. The grabenal basins of the Grand Banks region generally contain less than 7 km of section, but in places the pre-Cretaceous rocks may be 6 km thick in the basin centres. Five basins contain significant thicknesses of pre-Cretaceous sedimentary formations (named South Whale, Whale, Horseshoe, Carson and Jeanne d'Arc).

The oldest rocks encountered in deep offshore drilling of the younger basins on the East Canadian margin are found on the Grand Banks. Upper Triassic (Carnian-Norian) continental redbeds have been penetrated in a number of holes. They are overlain by an evaporitic and saliferous formation in excess of 2,000 m thick in places (*Argo Formation* and *Osprey evaporites*) which in age span the Triassic-Jurassic boundary. The redbed-salt sequence may initially have been influenced by horst-graben features, but deposition gradually spread out to cover wider areas. Dolomites of the *Iroquois Formation* (more than 500 m) follow upon the salt and in places extend out to rest on pre-Jurassic and basement rocks. There followed a period of regression (*Whale Unit*: Pliensbachian to Bathonian) during which firstly neritic shales and then shallow water carbonates were deposited across the Grand Banks.

A variety of facies related to outer-shelf carbonate bank complexes comprise the formation which abruptly overlies the Whale Unit (*Abenaki Formation*: 600 m, western Grand Banks). Coeval (Middle-Upper Jurassic) shales and subordinate sandstones occur on the Grand Banks and on the western flank of the East Newfoundland Basin and represent inner to middle shelf depositional conditions.

A period of uplift occurred in Early Cretaceous time which resulted in a profound change in sedimentation patterns. The Avalon Uplift became an elevated land area at this time separating the East Newfoundland Basin to the northeast, the Scotian Basin to the southwest and the Flemish Basin to the southeast. Coarse clastic and volcaniclastic deposits were shed westwards and eastwards on the flanks of the uplift in continental and deltaic facies (*Missisaupa Formation* and *Logan Canyon Formation*). The marine incursion which followed upon Early Cretaceous regression did not succeed in inundating the Avalon Uplift until Cenomanian time. The dominantly sandstone beds of the *Logan Canyon Formation* were followed by deepening shelf conditions and the deposition of glauconitic shales and siltstones of the *Dawson Canyon Formation* (Upper Cretaceous: maximum thickness on the western flank of the uplift 750 m). Distal southern areas of the Grand Banks were probably starved of sediment at this time. The succeeding chalk deposits (*Wyandot Formation*) are found only on the flanks of the uplift and are missing due to erosion or non-deposition over most of the Grand Banks, although equivalent shales may be present in some areas. The coarsening-upwards *Banquereau Formation* of Tertiary age completed deposition in the Grand Banks region following regional uplift in the Palaeocene, although it is thin over the northern part of the uplift.

Little is known concerning Mesozoic deposition in the adjoining *Flemish Basin* to the southeast. It is believed that a Triassic to Cretaceous section is present which may be up to 6 km thick.

Scotian Basin

The Scotian Basin[6, 13, 15–17] is the largest of the eastern Canadian Basins and there are probably more available published data than for any of the other basins. There is also a relatively complete Mesozoic and Tertiary succession. A number of sub-basins are recognized within the main basin (Fig. 10.5), together with intervening positive elements. The inner shelf is comprised of a positive element, the *Lahave Platform*, with its eastward extension south of Newfoundland, the *Canso Ridge*. A

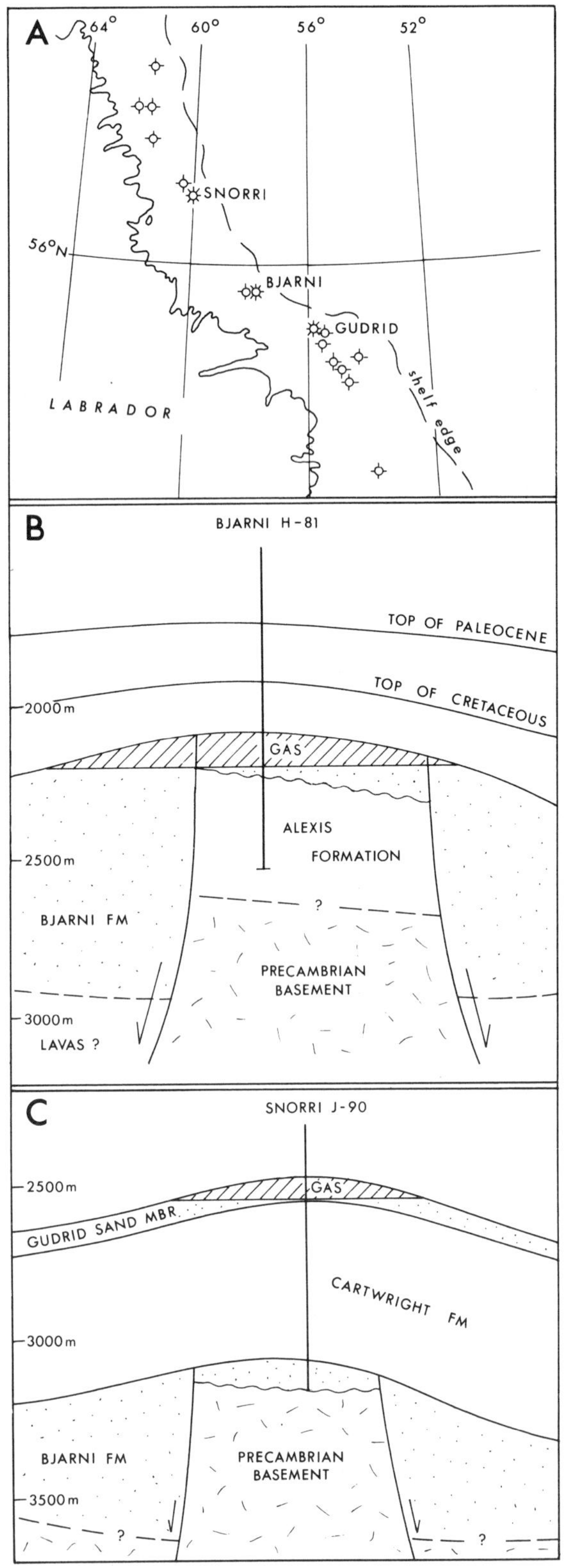

Figure 10.5. A. Location map of wells drilled on the Labrador Shelf showing three of the gas discoveries.
B. Diagrammatic cross-section of the Bjarni structure.
C. Diagrammatic cross-section of the Snorri structure.
The cross-sections with modification after Umpleby (1979).[4]

well defined hinge zone separates the platform and ridge in the north from the basinal area to the south. The basin contains up to 10–12 km of sediments and comprises a number of sub-basins. To the south again, in deep water more than 200 km from the present shore, a number of spectacular elongate salt piercements define a coast-parallel zone called the *Sedimentary Ridge Province.*

As in the other basins Mesozoic deposition began in the Scotian Basin with Triassic redbeds, which in this case are of Rhaetian age (*Eurydice Formation*). These beds have been penetrated by wells in the Orpheus Graben beneath a thick development of *Argo Formation* salt (Table 10.1). The original thickness of the basal redbed-salt formations may have been up to 3 km. The initial differentiation into horst and graben structures was probably gradually submerged beneath the spread of evaporites. The salt basin lay south and east of a hinge zone which formed the outer limit of the positive Lahave Platform. Redbeds interfinger with evaporites at the margin of the basin.

As in the East Newfoundland Basin widespread deposition of dolomite (*Iroquois Formation*) followed abrupt termination of salt deposition and this extended beyond the limits of the lower formations. Shallow marine carbonate conditions persisted from the Sinemurian to the end of Early Jurassic time. There then followed deposition of sandstones and variegated mudrocks which comprise the *Mohican Formation.* These sediments are of generally continental aspect and wedge out landward onto basement rocks. The upper contact with the overlying carbonates of the *Abenaki Formation* is gradational into cyclical oolites, reflecting the gradual onset of marine conditions in Middle Jurassic (Bathonian) time. Open marine conditions were established as the North American and African plates gradually separated. For the remainder of Middle and Late Jurassic time a well-defined pattern of sedimentation persisted on the Scotian Shelf; a carbonate bank and shelf carbonate complex with shallow water sandstones, shales and limestones to landward and basinal shales to seaward. Carbonate bank build-up took place on the outer shelf with a variety of platform interior carbonate facies on the landward side. Landward again of the carbonate bank were successively a deeper neritic zone with shalier facies, then nearshore sandier carbonate facies and lagoonal carbonates grading into shoreline clastics (Fig. 10.2). The correlative shallow marine to deltaic shales and sandstones of the *Mic Mac Formation* (up to 1.5 km thick) were deposited along the margin of the Scotian Basin. In the area of Sable Island, the Sable Sub-Basin deltaic build-up continued through much of the period. Seaward of the main carbonate platform variably calcareous marine basinal shales and minor sandstones or carbonates were deposited (*Verrill Canyon Formation*).

Regional subsidence and gradually deepening conditions led to the termination of carbonate deposition. Regressive clastic wedges began to prograde into the basinal areas. At this time the Sable Island delta prograded forward to fill first of all the Abenaki Sub-Basin and then the Sable Sub-Basin. The gradual progradation over the massive Abenaki carbonates means that the base of the clastics is strongly diachronous. By mid-Early Cretaceous time the carbonate shelf had been almost completely obliterated by the *Missisauga Formation* clastics.

The regression was associated with a major tectonic pulse which produced the Avalon Uplift and reactivated earlier structures. This also gave rise to rejuvenation of the hinterland from which large volumes of clastic sediments were then eroded. The extensive delta systems of the Missisauga and the succeeding *Logan Canyon Formation* locally exceeded 2.5 km thickness and persisted into the Cenomanian (Late Cretaceous). Only thin sandstone sequences occur on the positive Lehave Platform with thicker development in the Sable Island area. Offshore, in deeper water, deposition of prodelta shales and turbidites continued (part of the *Verrill Canyon Formation*).

A range of deltaic facies is present in the Missisauga Formation including top-delta coastal swamp (and coals), distributary and bay deposits. In Logan Canyon time a broad coastal plain environment is envisaged with many streams and a complex of coastal facies.

With deepening marine conditions in Late Cretaceous time the shale-dominant *Dawson Canyon Formation* was deposited. This unit is partly equivalent to the Logan Canyon Formation but gradually came to overlie it over much of the shelf.

Across the Scotian shelf the Dawson Canyon is usually 100–300 m thick and consists of shales with thin sandstones and siltstones. Shale deposition persisted until almost the end of Cretaceous time. However, there was a period of carbonate deposition in the Turonian (Petrel Limestone) and in latest Cretaceous and early Tertiary time carbonate (chalk) deposition dominated over wide areas (*Wyandot Formation*: Campanian-Maastrichtian). The Wyandot chalk was deposited on the floor of a deep water basin and formation thicknesses of the range 60–200 m are found. Landward (northwest) of the basin nearshore shales were deposited.

Coarser sediments entered the basin in the Palaeocene and began to fill the area of the Late Cretaceous chalk basin, while chalk deposition persisted further down slope. The *Banquereau Formation* (Tertiary) is 2 km thick on the outer Scotian Shelf and comprises a coarsening-upwards regressive sequence. The lower part of the succession represents sediment wedges prograding into the deep water basin. From Oligocene time onwards shallower water depths obtained, with complex submarine canyon systems eroding into the older shelf sediments.

Summary of Basin Development

For reasons of clarity each of the east coast basins have been described separately. However, from the accounts it will be clear that there is a measure of similarity between the basins, which were closely linked to the development of the early North Atlantic ocean. Early rift valley redbed sequences were followed by the Argo salt deposition which was initiated earlier in the central areas (Grand Banks) than in the marginal sub-basins. As in the Southern North Sea Basin, later diapiric movement of the mobile salt layer transformed the structural picture. Beneath the Scotian slope, in particular, elongate salt diapirs and walls, and shale masses occur in a zone called the Sedimentary Ridge Province (Fig. 10.2). Salt diapirism has been periodic in the several basins during the Cretaceous and Tertiary. In consequence stratigraphy may be locally complex around the episodically-moving structures, and a variety of structural closure is provided in the overlying sequences. The West Sable structure, for example (see below), has doming of Cretaceous sandstone reservoirs above a salt diapir possibly emplaced in three phases of salt movement.

Apart from halokinetic features the structural picture of the offshore east coast region is dominated by faulting. Tension and crustal thinning in the Triassic probably led to initiation of the well-defined rift systems. Late Triassic tectonism was accompanied by basaltic igneous activity. Regressive deposits about the Lower-Middle Jurassic boundary mark a further period of regional earth movements. An important phase of tectonic activity took place in latest Jurassic time, possibly due to final decoupling of the American and African plates. It was during this phase that the Avalon Uplift became clearly defined and considerably modified by block faulting, salt tectonism and erosion.

Uplift and erosion at mid-Cretaceous levels have been particularly important in determining the prospectivity of the Avalon Basin on the East Newfoundland Shelf. The basin contains a thick Lower Cretaceous section, in contrast to the Jeanne d'Arc Graben on the Grand Banks, and the beds were folded and partly eroded prior to deposition of the Upper Cretaceous strata.

The end product of the structural phases and salt tectonism described above was a number of distinct structural provinces. The Nova Scotian shelf, defined by northeast-trending faults, has been strongly affected by halokinesis. The Grand Banks is more obviously a fault-controlled province with narrow structural grabens, but salt structures are present within each of these. Further north the Labrador Shelf is dominated by fault-controlled coast-parallel basement ridges and troughs, although there has also been some salt tectonism. Deposition on the Labrador Shelf also began considerably later (Early Cretaceous) than in the southern basins.

In general terms each basin shows a thick wedge of clastic and carbonate sediments overlying strongly faulted basement. The major hinge zone faults have a history of continued growth and movement through Mesozoic and Tertiary times. Nevertheless, basement structure became gradually muted by the thick sediment cover and the Tertiary sediments show broad uniformity of facies.

Hydrocarbon Exploration and Potential

Offshore exploration on the East Canadian Shelf[9, 18] commenced in 1966. Since that time about 150 wells have been drilled, the majority of them on the Grand Banks and Scotian Shelf. There was an upsurge of exploration activity with a peak of drilling in 1973 followed by a slump to a low point in 1977. The reasons for the decline were the modest exploration success and high exploration costs.

Approximately half of the east coast wells have been drilled on the *Scotian Shelf*. The first drilling was on an offshore sand bar, called Sable Island, by Mobil, and located 250 m of net pay on the West Sable structure. However the follow-up drilling was less encouraging and the reserves are modest, mainly due to thin pay zones. The same is true of four other discoveries in the Sable Island area.[19] The West Sable structure itself is developed above a salt diapir. Dry gas, oil and oil/condensate zones have been located in sandstones (20–30% porosity) of the Dawson Canyon and Logan Canyon Formations of Cretaceous age, together with minor gas in the Jurassic. The salt diapirism which produced the faulted domal West Sable structure occurred in Late Cretaceous to Early Tertiary times. Discovery of further gas in the same area has led to consideration of development plans. One of these discoveries, the Cibralta structure northeast of Sable Island, had gas and condensate in massive sandstone units within the lower Mississauga and upper Mic Mac Formations (Fig. 10.3). The East Sable Venture D-23 well discovered a deep Lower Cretaceous gas play which may significantly increase the reserves of the area (now about 2 Tcf).

Drilling on the *Grand Banks* began in the early 1970s. Forty four wells have been drilled to date, with significant hydrocarbon shows only in the northeast. As we have seen earlier in the chapter the Grand Banks region is crossed by a number of grabenal basins. In the northeast there is a deep graben called the Jeanne d'Arc Sub-Basin which links into the *East Newfoundland Basin* (Avalon Basin). Wells in the early 1970s found promising hydrocarbon shows in Jurassic carbonates and sandstones and Cretaceous sandstones. The big break-through in East Canada offshore exploration occurred in 1979 when Chevron operating for a group on a Mobil farm-out block tested oil from three sandstone zones, in the Hibernia P-15 well within the Avalon Basin.[10]

Although the Hibernia Field lies on the outer part of the shelf, more than 300 km from the coast, the water depths are less than 100 m. The cross-section on Fig. 10.4 shows the field to be immediately east (seaward) of the main hinge zone fault, at the western margin of the Avalon Basin. There was considerable movement on the hinge zone fault during deposition – the prospective Lower Cretaceous section for example thickens towards the fault. At Lower Cretaceous level the Hibernia structure is a large rollover anticline which formed against the boundary fault, perhaps in response to flowage of salt at lower levels. Closure is afforded by the rollover against the fault to the west, basinward dip on the south and east, and by a series of northwest-trending faults in the north. The section beneath the mid-Cretaceous unconformity is folded and eroded, in contrast to uniform dips in the overlying Upper Cretaceous and Tertiary strata. The Tertiary thickens progressively eastwards from the hinge zone into the basin.

Traces of oil were found in early wells in the area but it was the P-15 discovery well and the O-35 and B-08 appraisal wells which defined the presence of a large oilfield. The pay zones (89 m net pay in P-15; 34 A.P.I. oil) are the Avalon, Hibernia and Jeanne d'Arc sandstones which have porosities averaging about 21%. An interesting feature of the structure is the variation in thickness of the Avalon Sandstone beneath the mid-Cretaceous unconformity – the unit has 136.5 m of net pay in the O-35 well, which is closer to the hinge zone, and only 17.1 m at P-15. Estimates of reserves in the Hibernia Field have been gradually increasing, but the current Newfoundland Government estimate is 1.8 billion barrels of oil with 2 Tcf of gas. Production average capacity has been estimated at 250,000 b/d.

The Ben Nevis structure 39 km southeast of Hibernia, drilled by Mobil in 1980, had 30 m of net hydrocarbon-bearing sandstone but is apparently uncommercial. However, drilling in 1981 on the Hebron structure, midway between Hibernia and Ben Nevis has been more successful in finding thicker pay section and Hebron appears likely to be commercial. Other structures are now being tested. The oil potential in the Hibernia region depends on a particular stratigraphic development in the Lower Cretaceous allied with a specific structural setting. The degree of growth on the controlling faults and the level of mid-Cretaceous erosion are critical factors in this equation. It remains to be seen over how large an area this combination of geological factors extends, but the Hibernia area already promises to be a major oil province.

The *Labrador Shelf*[6, 7] has been the site of oil exploration drilling since 1971. More than twenty wells have been drilled in this block-faulted province (Fig. 10.3). Large basement Palaeozoic fault blocks are draped by Mesozoic and Lower Tertiary sandstone reservoirs. A number of gas discoveries have been made in Palaeozoic dolomites, Lower Cretaceous and Tertiary sandstones, with Upper Cretaceous shales an important source and sealing horizon. Gas has been tested from a variety of horizons – Palaeozoic carbonate outliers above basement horsts (the Guldrig (Carboniferous) and Hopedale (Ordovician) wells), Lower Cretaceous fluvial sandstones draped over and abutting against basement structures (Bjarni well) and Lower Tertiary Palaeocene marine sandstones (Snorri discovery) or deltaic sandstones (Hekja well). The region is believed to have a high gas potential, probably due to the dominance of terrestrial organic matter in the pre-Tertiary section, and a relatively low potential for oil.

No oil has yet been discovered but in future it may be found in the deeper and more mature parts of the basin. At the present time there are no plans to develop the gas, but drilling is continuing. Drilling here is expensive, partly due to the water depths which range up to 1,000 m, but also because icebergs restrict the drilling season to 3–4 months and require that dynamically-positioned drilling rigs be used.

References

1. HAWORTH, R. T. 1975. The development of Atlantic Canada as a result of continental collision – evidence from offshore gravity and magnetic data: *in* Yorath *et al.* (*Eds*). Canada's Continental Margins and offshore petroleum exploration. *Mem. Can. Pet. Geol.* **4**, 59–77.
2. HAWORTH, R. T., GRANT, A. C. and FOLINSBEE, R. A. 1976b. Geology of the continental shelf off southeastern Labrador. *Pap. geol. Surv. Can.* **76–1C**, 61–70.
3. McMILLAN, N. J. 1973. Shelves of Labrador Sea and Baffin Bay, Canada: *in* McCrossan (*Ed.*) The Future Petroleum Provinces of Canada. *Mem. Can. Pet. Geol.* **1**, 473–517.
4. UMPLEBY, D. C. 1979. Geology of the Labrador Shelf. *Pap. geol. Surv. Can.* **79–13**, 34pp.
5. CUTT, B. J. and LAVING, J. C. 1977. Tectonic elements and geologic history of the South Labrador and Newfoundland continental shelf, eastern Canada. *Bull. Can. Pet. Geol.* **24**, 1037–1058.
6. PURCELL, L. P., UMPELBY, D. C. and WADE, J. A. 1980. Regional geology and hydrocarbon occurrences off the east coast of Canada: *in*

Miall (*Ed.*). Facts and principles of world petroleum occurrence. *Mem. Can. Pet. Geol.* **6**, 551–565.

7. McWHAE, J. R. H., ELIE, R., LAUGHTON, D. C. and GUNTHER, P. R. 1980. Stratigraphy and petroleum prospects of the Labrador Shelf. *Bull. Can. Pet. Geol.* **28**, 460–488.
8. JANSA, L. F. and WADE, J. A. 1975a. Geology of the continental margin off Nova Scotia and Newfoundland: *in* Offshore Geology of Eastern Canada; *Pap. geol. Surv. Can.* **74–30**, 2, 51–105.
9. THOMPSON, W. B. 1980. Recent exploration developments on the east coast of Canada. *Lecture to Canadian Gas Association Annual Meeting*, Jasper, Alberta.
10. McKENZIE, R. M. 1981. The Hibernian – a classic structure. *Oil and Gas Journal* **79** No. 38 Sept. 21, 1981. 240–246.
11. RUFFMAN, A. and HINTE, J. E. van. 1973. Orphan Knoll – a 'chip' off the North American 'plate'. Offshore Eastern Canada Symposium. *Pap. geol. Surv. Can.* **71–23**, 407–449.
12. BARSS, M. S., BUKAK, J. P. and WILLIAMS, G. L. 1979. Palynological zonation and correlation of sixty-seven wells, eastern Canada, *Pap. geol. Surv. Can.* **78–24**, 118 pp.
13. SHERWIN, D. F. 1973. Scotian Shelf and Grand Banks: *in* McCrossan (*Ed.*) The future petroleum provinces of Canada. *Mem. Can. Pet. Geol.* **1**, 519–559.
14. AMOCO, and IMPERIAL OIL, 1973. Regional geology of the Grand Banks, *Bull. Can. Pet. Geol.* **21**, 479–503.
15. McIVER, N. C. 1972. Cenozoic and Mesozoic stratigraphy of the Nova Scotia Shelf. *Can. J. Earth Sci.* **9**, 54–70.
16. GIVEN, M. M. 1977. Mesozoic and Early Cenozoic geology of offshore Nova Scotia. *Bull. Can. Pet. Geol.* **25**. 63–91.
17. WADE, J. A. 1977. Stratigraphy of Georges Bank Basin: Interpreted from seismic correlation to the western Scotian Shelf. *Can. J. Earth Sci.* **14**, 2274–2283.
18. SHERWIN, D. F. 1980. Geological setting: east coast offshore oil and gas. Lecture to '*Offshore Environment in the 80's Workshop*' St. John's, Newfoundland.
19. SMITH, H. A. 1975. Geology of the West Sable structure: *in* Yorath *et al.* (*Eds*). Canada's Continental Margins and offshore petroleum exploration. *Mem. Can. Pet. Geol.* **4**, 133–153.
20. WADE, J. A., GRANT, A. C., SANFORD, B. V. and BARSS, M. S. 1977. Basement structure Eastern Canada and adjacent areas. *Geol. Surv. Can. Map 1400 A* (Scale 1:2 million), 4 sheets.

Chapter 11

West Scottish Basins

Introduction

A chain of asymmetric sedimentary basins extends northwards from the Malin Sea through the Sea of Hebrides and the Minches, and northwards across the shelf lying to the west of the Orkney and Shetland Islands (Fig. 11.1). These are strongly fault-controlled basins containing often more than 5 km of sediments. A number of important faults or fault systems – the Rona Fault Belt, the West Shetland Boundary Fault, the West Minch Fault, the Skerryvore Fault and the Great Glen Fault – define the major structural units of northernmost Scotland. The basins are elongate in a caledonid northeast-southwest direction, probably related to earlier lines of weakness.

University-initiated regional geophysical surveys gave the first indications, in the 1960s, of thick sedimentary basins north and west of Scotland. This work was followed in the late 1960s by the first of the petroleum industry deep reflection seismic surveys. In the last decade many thousands of kilometres of high quality seismic reflection lines have been acquired in this region. The first deep drilling took place in 1972 in the West Shetland Basin, and all the drilling since that time – some 31 wells to the end of 1981 – has been in the same basin or on the immediately adjoining eastern margin of the Faeroe-Shetland Trough.

The *Shetland Platform* is a long-standing positive element between the North Sea and West Scottish sedimentary basins. The platform may be regarded as comprising the West Shetland and East Shetland offshore platforms, together with the intervening emergent mainland and island chain. Northeast of the Shetland Islands the West and East Shetland Platforms merge into a narrowing spur-like feature. The mainland coast of north Scotland, and the tiny offshore islands of Sule Skerry, Skerry Stack and North Rona, together with most of the Shetland Islands, are composed of Precambrian and Lower Palaeozoic basement rocks. The only other rock types exposed are Devonian (Old Red Sandstone) continentally-deposited sands and pebble beds. These occupy a northeast-trending tract (Orcadian Basin) from the Moray Firth, through Caithness and the Orkney Islands into the southern edge of the Shetland Islands (Fig. 11.2). Offshore sampling on the West Shetland Platform has shown that it consists of Precambrian and Lower Palaeozoic rocks together with considerable areas of Old Red Sandstone.

The *West Shetland Basin* is an elongate thick accumulation of Mesozoic strata northwest of the West Shetland Platform between latitudes 60° and 61° North. It is separated from the deep water Faeroe Basin to the west by the narrow basement uplift of the *Rona Ridge*. The basin is of the asymmetric trapdoor type (Fig. 11.2) with the thickest area of deposition in the southeast alongside the system of faults comprising the West Shetland Boundary Faults, which have vertical displacements of the order of 3,000 m. In the northwest the Rona Fault Belt Complex is the bounding fault system of the Rona Ridge into the

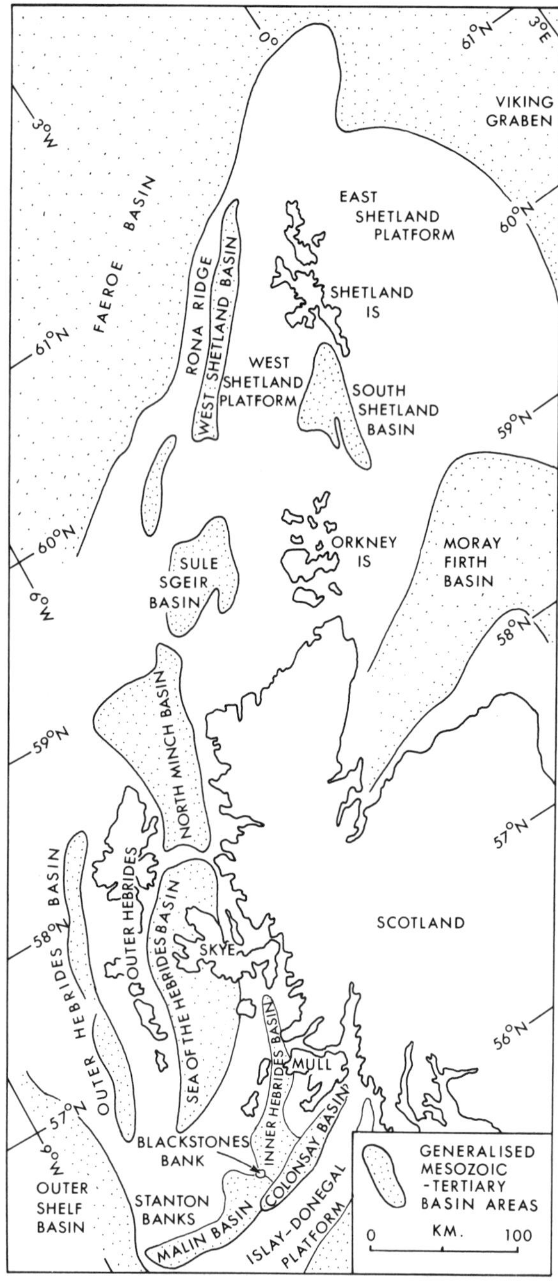

Figure 11.1. Generalized diagram showing the distribution of positive elements and offshore basins around the north and west of Scotland.

Faeroe-Basin and the faults display very large down-to-the-ocean displacements. The wedge of mainly Mesozoic strata which comprises the West Shetland Basin is itself broken by faults, some of which downthrow the east, affecting both basement and younger rocks and producing tilted fault blocks. The basin dies out southwards at about 60°N as the throw of the Boundary Fault diminishes.

To the west of the Outer Hebrides there is a narrow and probably shallow basin or basins, the *Outer Hebrides Basin* about which there is scanty information. In the Minch–Sea of Hebrides region, four small sedimentary basins have been downfaulted into the continental shelf. Two of these, the *North Minch Basin* and the *Sea of the Hebrides Basin* form a pair of elongate troughs closely paralleling the eastern coast of the Outer Hebridean Islands and separated from each other by a northwest-trending ridge of basement rocks (Fig. 11.3). The basins are bounded on the northwest by the major Minch Fault, against which the thickest zone of sediments (more than 1.4 km thick) is often found. The third and fourth basins in this region are the *Inner Hebrides Basin* and the minor *Colonsay Basin*, a pair of subparallel troughs to the southeast in the vicinity of Mull. The former is bounded on the northwest margins by the Skerryvore Fault trend and the latter on the southeast by the Great Glen Fault, both of which are known to have had a long and complex history of movement. The Inner Hebrides Basin terminates in the south against the basic layered intrusive centre of the Blackstairs Bank. The Malin Basin is the southward extension of the Inner Hebrides basinal trend beyond the Blackstones Bank (Fig. 11.3). To the south of the Great Glen Fault Line is the Islay-Donegal Platform, comprised of Lower Palaeozoic and Dalradian rocks, which forms a convenient divider between the West Scottish Basins described in this Chapter and those to the south between Southwest Scotland and Ireland which are dealt with in Chapter 12.

Geophysical Data

By the 1960s offshore gravity data were being correctly interpreted as being due to a series of fault-bounded sedimentary basins. Although the positions and form of the basins were only broadly known at that time, these original investigations served as a basis for all the subsequent more detailed geophysical investigations.[1–4] Seismic reflection data in conjunction with sea floor sampling have provided the means of refinement of the limits and depths of the basins. These surveys also provided some evidence for the age, thickness and nature of the infilling sediments, which in the West Shetland Basin have been further confirmed by deep drilling (only the outline results

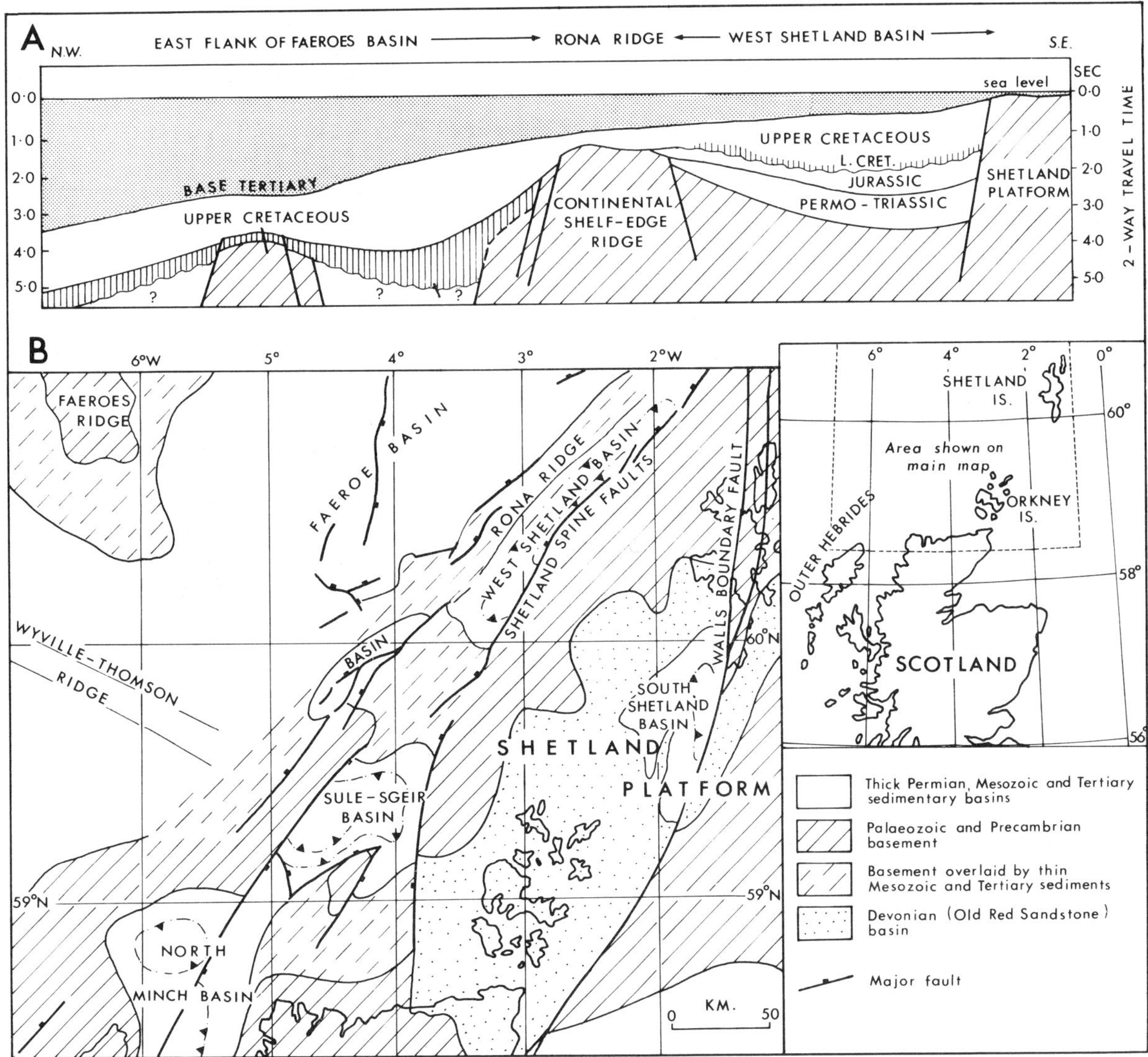

Figure 11.2 A. Section across the southern part of the West Shetland Basin based in part on a seismic profile.. (Modified after Naylor and Mounteney (1975)[11] and Fyfe, Abbots and Crosby (1981).[21])
B. Geological sketch map of the West Shetland area (in part after Ridd (1981)[14]).

of which have so far been released by the oil industry).

The gravity data north of Scotland showed a strong northeast-southwest belt of low gravity, lying west of the Shetland Islands, in the area of what is now termed the West Shetland Basin. To the west of the Orkney Islands a roughly circular zone of low gravity similarly outlined the small Sule Sgeir Basin (Fig. 11.1). This small basin is considered from geophysical evidence[6] to be fault-bounded on the northwest and to have a probable Mesozoic infill.[10, 11] However, it has also been depicted as a small Old Red Sandstone Basin.[12] On the West Shetland Platform immediately southwest of the Shetland Islands there is geophysical evidence[5] to suggest a further small relatively shallow (2 km) Mesozoic basin fault – bounded on its eastern side (Fig. 11.1).

Off the west coast of the Scottish mainland a steep gravity gradient follows the line of the Minch Fault, separating the high gravity basement rocks of the Outer Hebridean Islands from a region of marked gravity lows centred over Northern Skye and the Little Minch (Sea of the Hebrides Basin), and to the east and northeast of Stornaway, Lewis (North Minch Basin). To the southeast two elongate gravity

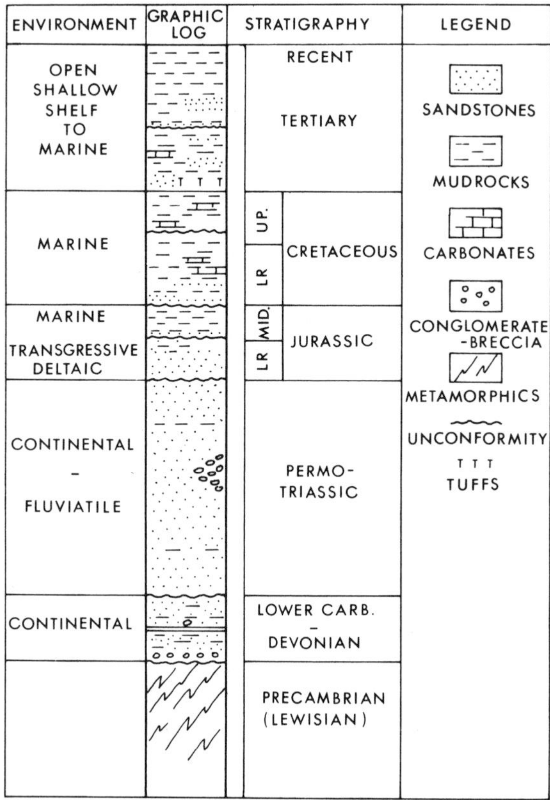

Figure 11.3. Stratigraphic summary of the West Shetland Basin (adapted from Jenkins and Twombley (1980),[15] with information from Ridd (1981)[14]).

lows are centred in the offshore region between Mull and the high gravity Coll-Tiree basement ridge (thus outlining the Inner Hebrides Basin), and in the vicinity of Colonsay (Colonsay Basin). The Blackstones igneous centre forms the most striking positive feature on the gravity maps of the southern part of the region. The Malin Basin is not distinguished by a well-defined gravity low, perhaps because of the general regional gravity rise towards the continental margin.

Interpretation of the regional magnetic anomalies[7] over the Sea of Hebrides and the Minches shows the marine area to be characterized by a pattern of almost rectangular anomalies separated by regions of steep gradient. These identify areas of volcanic activity within the Tertiary igneous province and, in general, are positive magnetic anomalies. Two broad regions of negative anomaly are centred over the gabbros and unltrabasic rocks of western Skye and western Mull. West of the Minch Fault there is an abrupt change in magnetic signature and the Lewisian basement complex of the Outer Hebrides is represented by anomalies of low magnetic relief, broken only by a chain of positive anomalies along the line of the Outer Isles Thrust. Further north the magnetic map[4] is a clear expression of the main elements of the Shetland Platform and basins.

To the south, in the Sea of Malin,[8] a complex magnetic picture emerges with volcanic centres and flows marked by anomalies; the Blackstones Bank igenous centre producing a strong circular negative anomaly. The wide basement areas separating the volcanic centres and sedimentary basins are again characterized by a quiet magnetic signature.

The faltering steps towards an understanding of the geology of the West Scottish offshore region, represented by the gravity and magnetic surveys, has been followed in all the basins by deep reflection seismic programmes undertaken both by the Institute of Geological Sciences[9] and oil companies. The West Shetland Basin, by reason of the greater density of seismic lines and the control afforded by deep drilling, is now known in some detail. In general the amount of seismic data and the confidence level in its interpretation falls off southwards in the group of basins. Nevertheless, the geophysical data, together with sea floor sampling and a knowledge of the Mesozoic stratigraphy of the emergent portions of the basins onshore, allows a reasonable interpretation of the seismic stratigraphy to be given.

Stratigraphy

Before turning to the stratigraphy of the basins listed above it would be advantageous to summarize the onshore geology, particularly that of the Mesozoic outcrops. West of Scotland, the Outer Hebridean Islands of Lewis, Harris, Uist and Benbecula are composed of highly metamorphosed Precambrian rocks similar to those forming the adjacent mainland of Scotland and parts of the Shetland Isles. However, throughout the Inner Hebridean Islands of Mull, Eigg, Skye and Raasay, and along the west coast of the mainland in areas such as Ardnamurchan, Morvern, Applecross and Gruinard Bay, small patches of Mesozoic rocks have been preserved either by downfaulting or as a protected sequence beneath the thick overlying layers of Tertiary lavas. These fragments represent the exposed margins of the offshore downfaulted basins, and as such provide valuable geological control for the age and lithology of the infilling sedimentary sequence.

Permian and Triassic sediments, containing potentially excellent reservoir horizons, are exposed as a sequence of coarse continentally-derived rocks comprising siltstones, sandstones and pebble beds. The sequence rests directly on the irregular basement floor of the troughs, and where exposed onshore can be seen to locally reach considerable thicknesses, e.g. 300 m on the east shore of the North Minch at Gruinard Bay, 4,000 m at Stornoway on Lewis and up to 200 m in South Mull. Marine limestones and sandstones of uppermost Triassic and basal Jurassic age are known only in the southern part of the area, in western Mull, where they are rapidly overstepped by younger Jurassic marine sediments. However, farther north, concurrent with the deposition of these marine strata, sediments were still being laid down in a continental environment.

Jurassic sediments attain a thickness of almost 1,000 m in the region of northern Skye, Raasay and Applecross where they are characterized by a thick series of sandstones, interbedded with hard, calcareous clays, and black oil-shales. The Middle Jurassic is locally dominated by a deltaic succession of shales, thin limestones and sandstones. Elsewhere, an incomplete Jurassic sequence is present on Mull and as minute exposures on the Isle of Arran, the Shiant Isles and on the mainland at Gruinard Bay.

A major break in sedimentary deposition occurred during the Late Jurassic and Early Cretaceous period throughout the Hebridean Province. By Upper Cretaceous times a thin (up to 20 m) but widespread sandstone and chalk sequence was deposited unconformly across the eroded surface of older Jurassic, Triassic and Permian rocks. Erosion has removed much of this Upper Cretaceous deposit, but a few scattered remnants have been preserved on Mull, Arran and around Loch Aline. These are composed of well-rounded, highly porous sandstones capped by a thin succession of chalk. Tertiary sediments appear to be absent throughout the Hebridean onshore region, although they are fairly extensively developed on the West Shetland Platform and in the Faeroe Basin. However, extensive areas of the Inner Hebridean Islands, and the floor of the Minches and Sea of Hebrides are covered by Tertiary igneous rocks. Marine sampling across the basins west of Scotland has yielded sea-floor cores of Triassic, Jurassic and Cretaceous age.

West Shetland Basin

This basin has seen the greatest amount of exploration activity on the west side of Britain and Ireland, and summary accounts of the geology have been released.[13–15] The basin comprises a thick Jurassic-Cretaceous sequence on the northwest of the West Shetland Platform and is separated from the Faeroe Basin by the Precambrian-Palaeozoic rocks of the *Rona Ridge* (Fig. 11.2). A system of *en echelon* faults (West Shetland Boundary Fault), including the Shetland Spine Fault, forms the southeast boundary against the platform, and the faults downthrow into the basin. The Shetland Spine Fault has a probable aggregate throw of 3,000 m. Southwards the throw of the boundary fault system decreases and older rocks crop out on the seabed. At its northeastern end the basin narrows, is broken by cross faults and contains more evidence of Tertiary igneous activity.

An outline of the stratigraphy and cross-sections of the basin are shown on Fig. 11.2. Wells in the vicinity of the Rona Ridge have penetrated fractured Precambrian (Grenvillian 970 ∓ 13 My) gneisses. The basement complex is unconformably overlain by a non-marine red bed sequence of Devonian to Tournaisian age. This comprises a sequence of interbedded pebble conglomerates, sandstones, mudrocks and cornstones. A wide range of alluvial plain, channel fill, point bar and flood basin environments are represented. These rocks are typical of the Old Red Sandstone sequence known from outcrops in the Orkneys and on the Scottish mainland. The onshore successions are chiefly unmetamorphosed sandstones with conglomeratic horizons, deposited in alluvial and lacustine environments. The whole Old Red Sandstone complex of North Scotland-Shetland appears to have been deposited in irregularly subsiding fault-controlled valleys with large lakes and dominantly eastward or northward flowing rivers.

The Old Red Sandstone of the West Shetland Basin (which continues into the Carboniferous in redbed facies, as in the Midland Valley of Scotland and in Ireland) is known to be more than 500 m thick on the Rona Ridge but thins to zero in places under the influence of an irregular depositional surface and later (pre-Cretaceous) fault movement. Greater thicknesses may occur in the deeper undrilled parts of the basin.

A thick (more than 1,000 m) dominantly red sequence of presumed Permo-Triassic age is preserved in the deeper southeast portion of the basin against the West Shetland Boundary Faults. It is dominantly a succession of red friable sandstones with subordinate mudstones and, near the Shetland Fault, local developments of breccia. Alluvial fan environments are envisaged, produced by rivers flowing westwards from the Shetland Platform across the active boundary fault systems. Tilted Basement and Old Red Sandstone blocks form an irregular surface beneath the unconformable younger cover. The whole sequence thins towards the Rona Ridge, where it is absent (Fig. 11.2).

A period of uplift and erosion preceeded deposition of the Jurassic; the Lower Jurassic is not represented due either to non-deposition or erosion. The ensuing Middle Jurassic transgression produced a thin (maximum 100 m) sheet of deltaic and marginal marine sandstones across the basin and these grade laterally and upwards into marine shales.

The widespread but thin (50 m) Upper Jurassic is typically in a dark, carbonaceous pyritic shale (Kimmeridge Clay) facies, locally with thin sandstone or siltstone bands. It clearly represents a period of maximum transgression within the tectonically active basin. Fault movement and regression culminated in the Late Jurassic to produce the Late Cimmerian erosional unconformity. The Jurassic was deeply eroded, particularly on tilted fault blocks, and the Jurassic as a whole is better preserved towards the southwest of the West Shetland Basin and Rona Ridge.

The unconformably overlying Lower Cretaceous is similarly confined to the southwest of the basin. It comprises marginal shelly glauconitic sandstones interbedded with, and interfingering westwards into, thinner marine mudstones and limestones. Successively younger stages overlap onto the Rona Ridge and the whole sequence is overstepped by Campanian-Maastrichtian beds on the higher parts of the ridge. The Lower Cretaceous in the main part of the basin is of the order of only 100 m thick. A similar facies distribution continued into the Cenomanian to Santonian (Upper Cretaceous) stages. Rocks of this age are mainly absent over the Rona Ridge, but following an intra-Upper Cretaceous unconformity open marine mudstones spread across the basin and over the ridge. This Campanian-Maastrichtian sequence comprises open marine mudrocks with thin limestone beds and represents a period of maximum marine transgression and tectonic calm. There was probably a minor amount of volcanism in this region towards the close of Cretaceous time. The Upper Cretaceous interval thickens westwards into the Faeroes Basin which was rapidly subsiding during this period.

Unlike the Mesozoic rocks, the Tertiary sediments form a northwestward thickening wedge which does not show the influence of older structural configurations (Fig. 11.2). The basal unconformity is followed by Lower Palaeocene mudrocks with reworked Jurassic microfossils indicative of widespread erosion of the underlying Jurassic. Tuffs, also widespread in the Northern North Sea, glauconitic sandstones and minor limestones are encountered. The Palaeocene is succeeded by a dominantly argillaceous sequence, within which there is a well-defined mid-Tertiary unconformity.

The stratigraphy of the flank of the *Faeroe Basin* adjacent to the Rona Ridge is known from seismic surveys and from a number of wells. There has been no drilling in the deeper parts of the basin. The Jurassic-Tertiary successions encountered on the northwestern flank of the Rona Ridge are generally thicker than the West Shetland Basin counterparts. The Middle Jurassic is represented by several hundreds of metres of dark grey marine shales with sandstones, and the Upper Jurassic shales are also much thicker and contain sandstone and conglomerate intercalations. The Lower Cretaceous of this zone, in contrast to the thin or eroded sequences of the Rona Ridge, is more than 1,000 m thick and comprises sandstones with an increasing conglomeratic component towards the base. This is overlain by more than 2,000 m of grey mudrocks, with thin sandstones, belonging to the Upper Cretaceous. The steep depositional slopes combined with erosion on the adjacent ridge gave rise to slumping and turbidites in this interval. The Tertiary thickens rapidly northwestwards from the ridge. A thickening prograding wedge is interpreted from the seismic data and this is overlain by a complex Oligocene to Recent sequence.

Sule Sgeir-South Shetland Basin

At the southwestern end of the West Shetland Basin the throw decreases on the Shetland Fault (southeastern boundary fault). The main down-to-the-basin-throw is then taken up by another fault *en echelon* to the west, to produce a narrower subsidiary basin in line with the Rona Ridge. This basin in turn terminates at about 59° 30′N. The Rona Ridge is replaced southwards by a series of fault blocks and the main Rona Fault Belt swings westwards. This confused picture is further complicated, at least in geophysical terms, by the circular gravity anomaly over the Sule Sgeir Basin.

Little is known concerning this basin.[1] Towards the west the basin narrows against the relatively shallow basement ridge along the northern extension of the Minch Fault. More than 3 km of sediments are anticipated close to the fault and those thin eastwards onto the West Shetland Platform. The basin fill probably comprises Tertiary sediments resting unconformably on a dipping older sedimentary sequence of Mesozoic or possibly Old Red Sandstone age.

South of the Shetland Islands is a further shallow basin[5] containing about 2 km of Mesozoic sediments which rest on older basement and Old Red Sandstone strata. The eastern boundary of this basin appear to be formed by a possible splay of the Great Glen Fault.

North Minch Basin

To the southwest of the relatively positive area comprised by the southern part of the West Shetland Platform between 59° and 59° 30′ lies the North Minch Basin. In its southern part the North Minch Basin[16] is constrained between the Minch Fault on the west and the Precambrian rocks of the mainland shelf on the east. To the north of the Outer Hebrides the basin changes trend and broadens to straddle the line of the Minch Fault. The basin has received less geophysical coverage than the basins to the north and south. The sedimentary fill is probably in the order of 4 km thick. It seems probable that the major part of this sequence comprises Permo-Triassic redbeds comparable to the thick Stornaway Beds which rest on the Lewisian platform west of the Minch Fault.[17] Middle Jurassic marine sediments have been sampled and may be overlain by thin Cretaceous or Tertiary strata; however sample control for the seismic stratigraphic interpretation is poor (Fig. 11.4). The Mesozoic strata everywhere rest unconformably on Precambrian crystalline gneisses and sediments.

Sea of Hebrides Basin

The Sea of Hebrides Basin is an extensive trough extending southwards from the Isle of Lewis (57° 55′N) beneath the Tertiary lavas of Skye and the Little Minch into the Sea of Hebrides, where it finally narrows to pinch out some 30 km south of Bernerary, Outer Hebrides (56° 60′N; Fig. 11.3). To the northwest the basin margin closely follows the coastline of the Outer Hebrides, although the main thickness of infilling sediments lies immediately east of the Minch Fault. At its southeast margin the basin floor gently shelves onto the Precambrian and Palaeozoic ridge of Rhum, Coll and Tiree. At the northeastern end of the basin the sediments abut sharply against a northwest-trending ridge of Torridonian (Late Precambrian) sediments which acts as a barrier between this basin and the North Minch Basin. The basin is probably relatively shallow north of Skye and the sediments dip radially towards the island. South of Skye the basin axis terminates southwards at an acute angle against the Minch Fault.

Permo-Triassic mudstone and Rhaetian to Hettangian (Triassic-Jurassic boundary) limestone have been recognized in two cores taken west of the Minch Fault.[18] These sedimentary units appear to be older than a reflector which can be traced seismically into the basin east of the fault, and which in turn lies beneath a well bedded, sequence which laterally equates with Bathonian (Middle Jurassic) sediments in western Skye. The main basin fill is therefore interpreted as being of Permo-Triassic to Jurassic in age. This sequence can be seen on seismic records to be capped by extensive lava flows. The total thickness of sedimentary and volcanic strata within the basin is in the order of 2.5 to 3.0 km. To the southeast of Skye 1 km of sediments of probable Palaeocene age occupy a small basin resting on downwarped Eocene basaltic lava sheets.[19]

To the west of the Outer Hebrides (Fig. 11.3) there is an elongate narrow and shallow basin (*Outer Hebrides Basin*) is developed along the western side of the Outer Hebrides Fault. Between 1 km and 2 km of Mesozoic strata may be present but little detail is known.[10, 11, 20] Gravity data suggest that the basin is deeper in the north adjacent to the Isles of Lewis and Harris and shallows southwards towards Barra and South Uist. The basin appears to be asymmetric with the deeper part against the faulted eastern margin, that is in the opposed sense to that of the basins in the Minches and Sea of Hebrides.

Inner Hebrides Basin – Malin Basin

The Inner Hebrides basin lies between the Skerryvore Fault in the northwest and the positive elements represented by the Iona Platform, Mull and the Ardnamurchan-Morar mainland coastline to the southeast. The basin is narrower than the Sea of Hebrides Basin although gravity and seismic reflection data show the maximum thickness of infilling sedi-

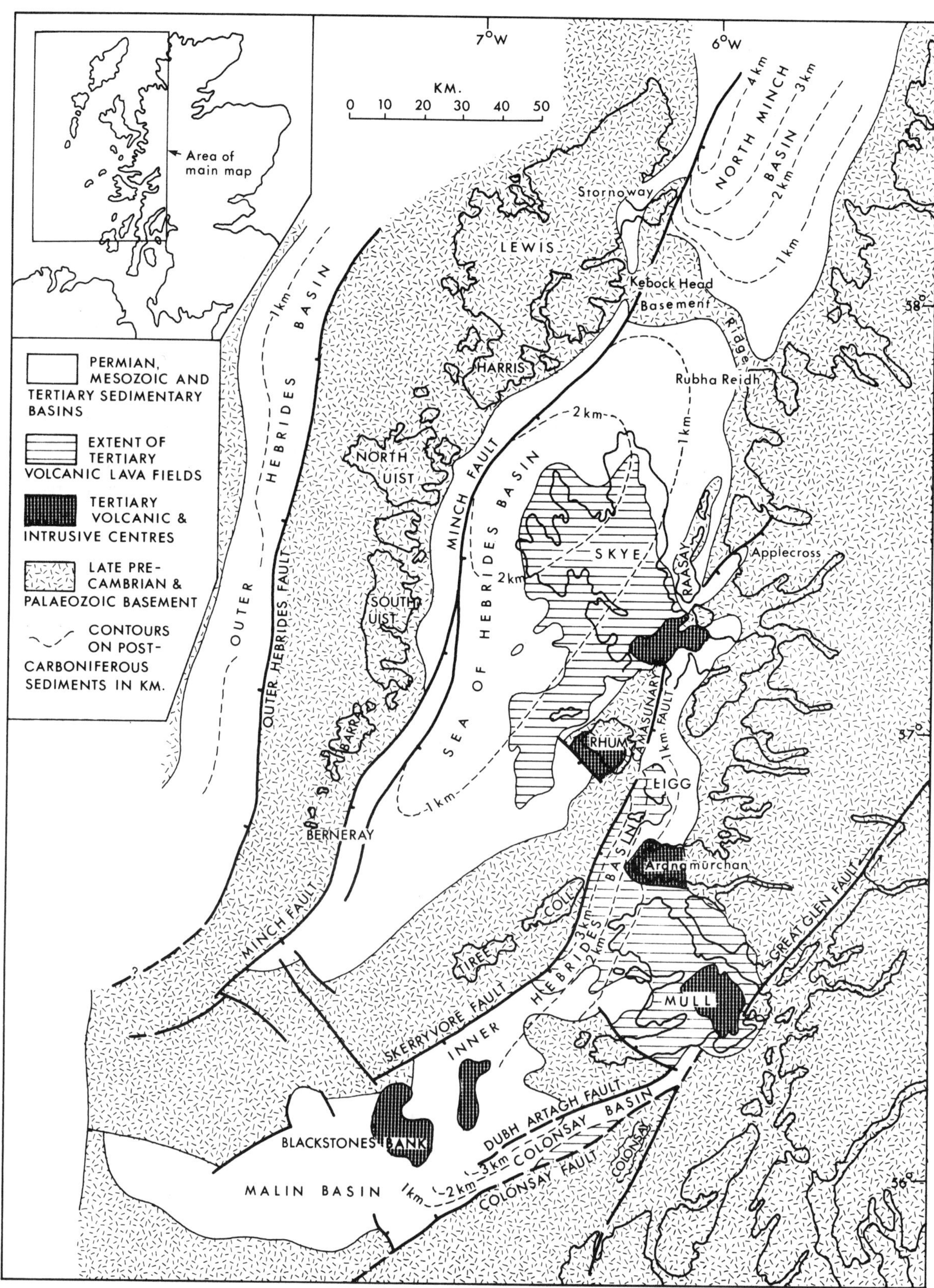

Figure 11.4. Geological sketch map of West Scotland and the Hebrides showing the offshore basins. (From Naylor and Mounteney (1975)[11] with modifications after Evans *et al.* (1980)[8] and Binns *et al.* (1975)[16]).

ments to be of similar order. Likewise the deepest part of the basin is located along its western margin adjacent to the boundary fault. Eastwards the sediments thin and overlap across the Palaeozoic basement rocks of Mull and the mainland coast (Fig. 11.4). The eastern margin is poorly defined, with overlap and faulted boundaries recognized in different places.

There is a thick Quaternary cover over much of the Inner Hebrides Basin which has hampered shallow seismic and sampling investigations. From the limited amount of deep seismic profiling it appears that relatively thick slightly folded Tertiary sediments of possible Oligocene age underlie the Quaternary unconformity. Little is know of the Mesozoic sequence which infills the trough. It is thought that the sediments will be similar to the Triassic sandstone and pebble beds exposed on the Isle of Inch, off western Mull, and to the Lower Jurassic calcareous sandstones, limestones and shales exposed in southern Mull. In the vicinity of Mull the thickest part of the trough is overlain by Tertiary lavas (Fig. 11.4).

The Camasunary Fault shows at least one phase of pre-lava movement and on Raasay a post-lava throw of at least 760m. The Camasunary Fault of Skye was formerly thought to extend southwards to Skerryvore southeast of Tiree.[18] More recent work however[8] has shown the marine fault to be an independent structure termed the Skerryvore Fault, with no contact with the Camasunary Fault which ends just south of Skye. The Skerryvore Fault has also undergone Tertiary and probably Mesozoic movement.

The southwestern limit of the Inner Hebrides Basin is arbitrarily taken at the Blackstones Bank. This layered basic intrusion has yielded a Cretaceous age of 84 million years and has a complex form, cylindrical in the main but sill-like in the north and related to smaller flows or intrusions in the southeast. Hornfelsed metasediments have been recovered from the neighbourhood of the bank.

The westward extension of the Inner Hebrides Trough beyond the Blackstones Bank is termed the *Malin Basin*. The basin trend hereabouts is more nearly westerly, with the Islay-Donegal Platform to the south and the Stanton Banks to the north. Faulting is evident on seismic records around the southern margin of the Stanton Banks. Little is known concerning the southern margin which may be in part a normal overstep of the basin fill onto the basement but which is partially obscured by an extensive lava field (Riddihough's Lava Field). At its western limit the Malin Basin plunges out at the shelf edge after breaking across the basement lip to the outer shelf.

The central portion of the basin is broadly folded whereas the southern and western margins are areas of complex faulting. The position of the Malin Basin on the outer shelf margin gives it a more complex structural setting than the faulted half-grabens of the Scottish margin further north. Seismic profiles show the existence of numerous minor intrusions and sills. As in the Inner Hebrides Basin there is a thick cover of Quaternary clay, frequently more than 100 m thick. A shallow borehole in the north of the basin, west of the Blackstones complex, penetrated ammonite-bearing siltstones which micropalaeontological evidence indicates to be Early to Middle Jurassic in age.

Colonsay Basin

The last of the West Scottish basins considered in this chapter is a narrow structural trough lying west of the Island of Colonsay (Fig. 11.3). The Colonsay Basin is controlled by the Great Glen Fault System, the main arm of which, the Dubh Artach Fault, forms the northwestern boundary, while an associated branch known as the Colonsay Fault forms the southeast margin. The Islay-Donegal Platform, of which Colonsay island is a part, is the major positive element on the south of the Colonsay Basin, dividing it from the complex of sedimentary basins lying between the northern coast of Ireland and Jura-Kintyre coastline of the Scottish mainland, which form the subject matter of the next chapter.

Thick Quaternary infill of a deeply eroded surface characterizes the Colonsay Basin. Shallow seismic profiling and gravity programmes have revealed the form of the basin, which exhibits a marked gravity gradient along its southeastern margin. Deep seismic information has been published only for the southwestern areas where the Quaternary grades down into uniform Tertiary (possibly Oligocene) sediments to a combined maximum thickness of 400 m. A strong reflector at this level may represent thin Tertiary basalts beneath which 1,200 m of layered sediments are thought to be of Mesozoic age. Well bedded strata of similar appearance are seen at the basin margins on shallow sparker profiles and have yielded a sample of Middle Jurassic age.[8]

Although the Dubh Artach Fault is a seaward continuation of the Great Glen Fault Line it is less well defined than the Colonsay

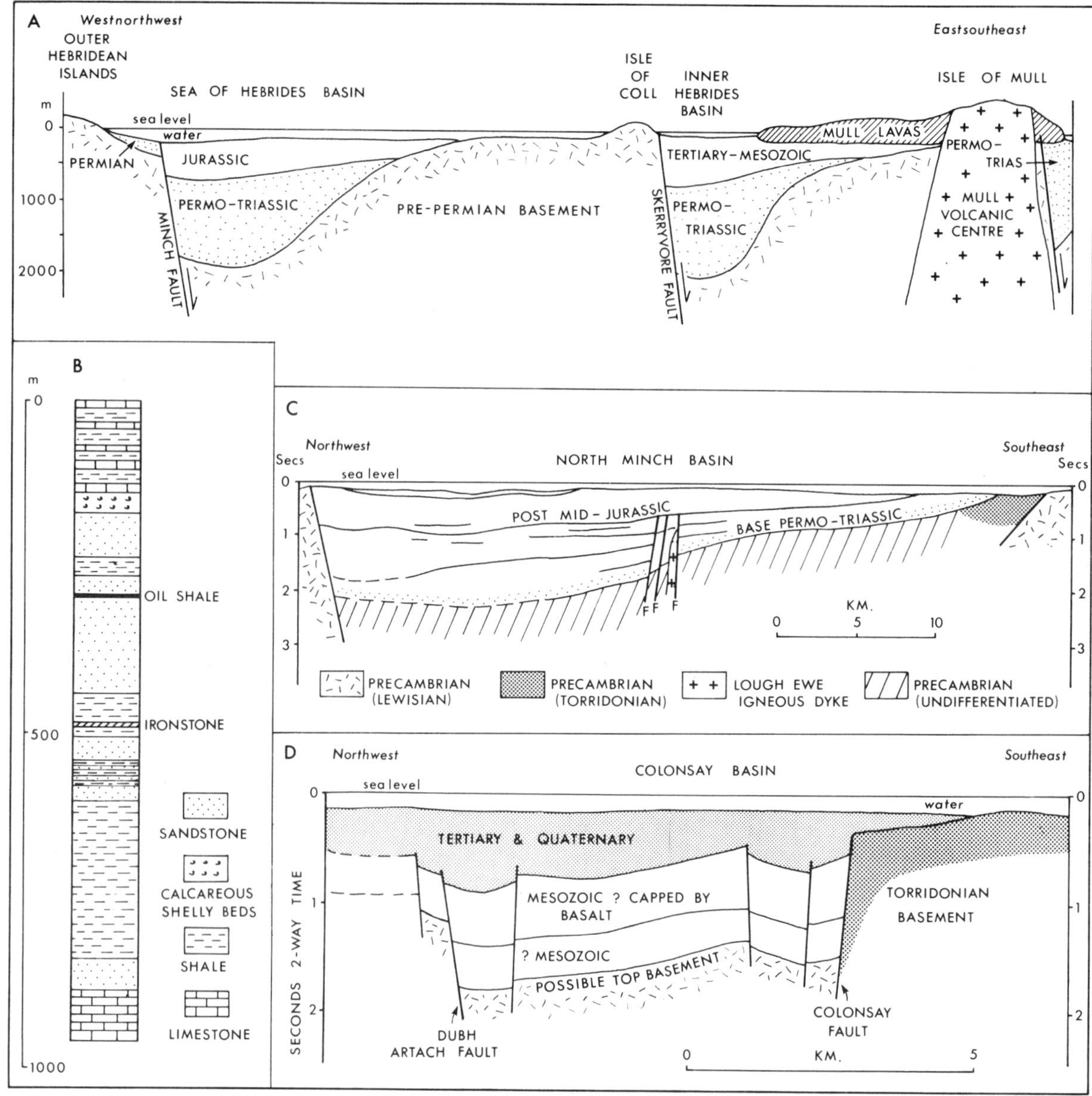

Figure 11.5A. Diagrammatic cross-section across the Sea of Hebrides and Inner Hebrides Basins (modified after Naylor and Mounteney (1975).)[11]
B. Typical stratigraphic sequence of the Jurassic strata in the Inner Hebrides.
C. Section across southern part of the North Minch Basin based on seismic profiles (adapted from Binns *et al.* (1975)[16]).
D. Section across the Colonsay Basin based on a deep seismic reflection profile (adapted from Evans *et al.* (1980)[8] Figure 16).

Fault on seismic profiles or magnetic and gravity maps. The Dubh Artach Fault is not clearly seen southwest of the cross fault which forms the southwest limit to the shallowing basin. The bounding faults on both margins of the basin are stepped and compound in part. There is no great asymmetry to the basin, although the northern part contains somewhat the thicker sediment fill. Along the southwestern portion of the northern margin Mesozoic strata extend across the boundary fault rim and into the adjoining Malin Basin. Nearer to the mainland the Iona Platform is a complex of Lewisian, Torridonian and Moinian basement rocks, igneous rocks, with a cover of Permo-Triassic sandstones. Colonsay and the surrounding shelf area of Torridonian rocks south of the basin are a part of the Islay-Donegal Platform. This feature, comprised of Lower Palaeozoic and Upper Precambrian rocks extends southwestwards along the trend of the Great Glen Fault Line towards the north coast of Ireland, and is penetrated in places by Tertiary intrusions.

Hydrocarbon Exploration and Potential

Oil and gas shows were encountered in several of the early wells drilled in the West Shetland Basin; at Basement, Jurassic and Lower Cretaceous levels.[14] In 1977 B.P. discovered a major accumulation of heavy (22–25° API) oil in Block 206/8 (Fig. 11.2) on the Rona Ridge. Oil-in-place figures of 3.5–4.0 billion barrels have been quoted, but reservoir quality and complexity combined with the oil viscosity pose problems for field development. The oil is contained in fractured Precambrian rocks and in the variable porosity of the Old Red Sandstone sealed by overstepping Upper Cretaceous mudstones. It has been postulated that the thicker, and deeper, Jurassic or Cretaceous source rocks of the Faeroe Basin are more likely to have produced the oil than their equivalents to the east. The depth of burial and timing of migration are critical factors in this equation. Although untested structures remain in the West Shetland Basin the main focus of exploration activity may move into the Faeroe Trough. Nevertheless unconformity subcrop possibilities form potential untested traps at both Old Red Sandstone and Permo-Triassic levels, as do basin margin clastic wedges against the boundary faults together with Cretaceous onlap traps.

The sandstones of the Old Red Sandstone may be considered as potential target reservoirs throughout the North Scotland-Shetland province. Their reservoir properties are variable but sometimes reasonably good, and sandstone thicknesses are substantial. In structural highs they could provide adequate reservoir capacity for oil.

Other reservoir horizons are the clastic developments within the Permo-Triassic and Middle Jurassic, and possibly Lower Cretaceous sandstones and limestones. The main potential traps are fault- and unconformity-related. Tilted fault blocks and subcropping units beneath unconformity surfaces are well demonstrated in the West Shetland Basin, as are Cretaceous onlap possibilities. Good hydrocarbon source rocks are present in the Cretaceous, particularly the Kimmeridgian shale. The main problems are related to finding structures and reservoirs adjacent to mature source rocks. Many potential source horizons in the West Shetland Basin have probably not been sufficiently deeply buried to achieve maturity. To the west the potential of the deeper water Faeroe Basin remains virtually unknown and untested. It seems likely, however, that mature source rocks will be found there.

The shallower and smaller West Scottish basins south of the West Shetland Basin must be regarded as having a significantly lower potential. As a result they remain untested by deep drilling and have been covered by a lower density of geophysical surveys. Lack of mature source rocks probably represents the main problem, although the outer shelf basins must be regarded as having some prospectivity. The thick reservoir sequences frequently lack adequate cap rocks.

References

1. BOTT, M. H. P. and WATTS, A. B. 1970. Deep seidmentary basins proved in the Shetland-Hebridean continental shelf and margin. *Nature (London)* 225, 265–268.
2. BOTT, M. J. P. and WATTS, A. B. 1971. Deep structure of the continental margin adjacent to the British Isles: *in* Delaney, F. M. (*Ed.*). The Geology of the East Atlantic continental margin, *2. Inst. geol. Sci. Rep.* 70/14, 89–109.
3. WATTS, A. B. 1971. Geophysical investigations on the continental shelf and slope north of Scotland. *Scott. Jl. geol.* **7**, 189–218.
4. FLINN, D. 1969. A geological interpretation of the aeromagnetic maps of the continental shelf around Orkney and Scotland. *Geol. Jl.* **6**, 279–92.
5. BOTT, M. H. P. and BROWITT, C. W. A. 1975. Interpretation of geophysical observations between the Orkney and Shetland Islands. *Jl. geol. Soc. Lond.* **131**, 353–371.
6. BOTT, M. H. P. 1975. Structure and evolution of the North Scottish Shelf, the Faeroe Block and the intervening region: *in* Woodland, A. W. (*Ed.*). *Petroleum and the Continental Shelf of North-West Europe 1*, Applied Science Publ., London. 105–113.
7. BULLERWELL, W. 1964. Aeromagnetic map of part of Great Britain and Northern Ireland.

Sheets 7–9, *geol. Surv. U.K.*
8. EVANS, D., KENOLTY, N., DOBSON, M. R. and WHITTINGTON, R. J. 1980. The geology of the Malin Sea. *Rep. Inst. geol. Sci. London* 79/15, 44pp.
9. McQUILLIN, R. and BACON, M. 1974. Preliminary report on seismic reflection surveys in the seas around Scotland, 1969–73. *Rep. Inst. geol. Sci. London* 74/12, 7pp.
10. WHITBREAD, D. R. 1975. Geology and petroleum possibilities west of the United Kingdom: *in* Woodland, A. W. (*Ed.*). *Petroluem and the Continental Shelf of North-West Europe, 1*, Applied Science Publ. London, 45–59.
11. NAYLOR, D. and MOUNTENEY, S. N. 1975. *Geology of the Northwest European Continental Shelf* Vol. 1. Graham, Trotman & Dudley London, 162 pp.
12. KENT, P. E. 1975. The tectonic development of Great Britain and the surrounding seas: *in* Woodland, A. W. (*Ed.*). *Petroleum and the Continental Shelf of North-West Europe, 1*, Applied Science Publ. London, 3–28.
13. CASHION, W. W. 1975. The geology of the West Shetland Basin. *Proc. Offshore Europe Conf. 1975*, Spearhead Publ., Kingston-on-Thames, 216.1–216.7.
14. RIDD, M. F. 1981. Petroleum geology West of the Shetlands: *in* Illing, L. V. and Hobson, G. D. (*Eds*), *Petroleum geology of the Continental Shelf of North-West Europe.* Heyden & Son Ltd, London, 414–425.
15. JENKINS, D. A. L. and TWOMBLEY, B. N. 1980. Review of the petroleum geology of offshore northwest Europe. *Trans. Instn. Ming. Metall.* Special Issue: Petroleum, 6–23.
16. BINNS, P. E., McQUILLIN, R., FANNIN, N. G. T., KENOLTY, N. and ARDUS, D. A. 1975. Structure and stratigraphy of seidmentary basins in the Sea of Hebrides and the Minches: *in* Woodland, A. W. (*Ed.*). *Petroleum and the Continental Shelf of North-West Europe Vol. 1*. Applied Science Pub., London, 93–102.
17. STEEL, R. J. and WILSON, A. C. 1975. Sedimentation and tectonism (? Permo-Triassic) on the margin of the North Minch Basin, Lewis. *J. geol. Soc. London* **131**, 183–202.
18. BINNS, P. E., McQUILLIN, R. and KENOLTY, N. 1974. The geology of the Sea of Hebrides. *Rep. Inst. Geol. Sci.* 73/14, 43pp.
19. SMYTHE, D. K. and KENOLTY, N. Tertiary sediments in the Sea of Hebrides. *J. geol. Soc. London* **131**, 227–233.
20. JONES, E. J. W. 1978. Seismic evidence for sedimentary troughs of Mesozoic age on the Hebridean continental margin. *Nature* 272, 789–792.
21. FYFE, J. A., ABBOTTS, I. and CROSBY, A. 1981. The subcrop of the Mid-Mesozoic unconformity in the U.K. area: *in* Illing, L. V. and Hobson, G. D. (*Eds*). *Petroleum Geology of the Continental Shelf of North-West Europe*, Heyden & Son Ltd, London, 236–244.

Chapter 12

Northern Ireland-Southwest Scotland Basins

Introduction

As we have seen in the previous chapter a complex of fault-controlled Mesozoic basins extends from the Shetlands along the west coast of Scotland. A similar development of basins occurs between southwest Scotland and the north coast of Ireland, and again between Britain and Ireland in the Irish Sea. For convenience of description an arbitrary limit to the West Scottish basins was drawn south of the Great Glen Line at the Islay-Donegal Platform. In this chapter we will be considering the basins south from the Islay-Donegal Platform to the northern margin of the Solway Firth Basin. These basins again have a dominantly northeast-southwest trend controlled by the major faults which frame the Midland Valley of Scotland and its analogue in Ireland. A glance at Fig. 12.1B will serve to demonstrate the dominant role played by faults in this region such that many of the geological boundaries on the map are defined by faults.

Although the surface extent of Mesozoic and younger sediments along the north coast of Ireland takes the form of a broad belt stretching from Strangford Loch northwestwards towards Loch Foyle, the major accumulations of sediments are within two fault-bounded troughs, the *Magee Basin* with its northward extension the *North Channel Basin*, and the *Rathlin Trough* (Fig. 12.1A). Taken together these basinal developments have been called the *Ulster Basin*. The two basins are separated by an intervening ridge of Dalradian metamorphic rocks, sometimes termed the *Highland Border Ridge*, which contains complex areas of Carboniferous, Mesozoic and Tertiary igneous rocks. This ridge, along with considerable portions of the onshore parts of the basins, is partly obscured beneath a thick cover of Tertiary plateau basalts. The basalt flows have produced the familiar Antrim scenery of the Giant's Causeway coast. The ridge is actually a structural extension of the Dalradian basement seen on the Mull of Kintyre and is bounded on its north flank by the Tow Valley Fault (Fig. 12.1). To the south the North Channel – Magee Basins extend southwards to Larne and Belfast, and towards the line of the Southern Upland Boundary Fault. An eastern limb of the basin extends into the Firth of Clyde as a submerged sedimentary trough. Further north, the Rathlin Trough is contained between the Tow Valley Fault to the southeast and the Foyle Fault in the northwest. The trough extends offshore northeastwards towards Scotland into the southern part of the Sound of Jura, between Islay and Kintyre.

Northwest of the Foyle Fault is a further Dalradian ridge which extends from Inishowen on the north Irish coast through the Middle Bank to Islay and Jura in Scotland. The northwest boundary of this element against the Islay-Donegal Platform (Fig. 12.1) is the Leannan-Loch Gruinart Fault. A narrow basin, the *Loch Indaal Basin*,[1] is preserved along the southeast side of the fault in the offshore, but is not represented onshore.

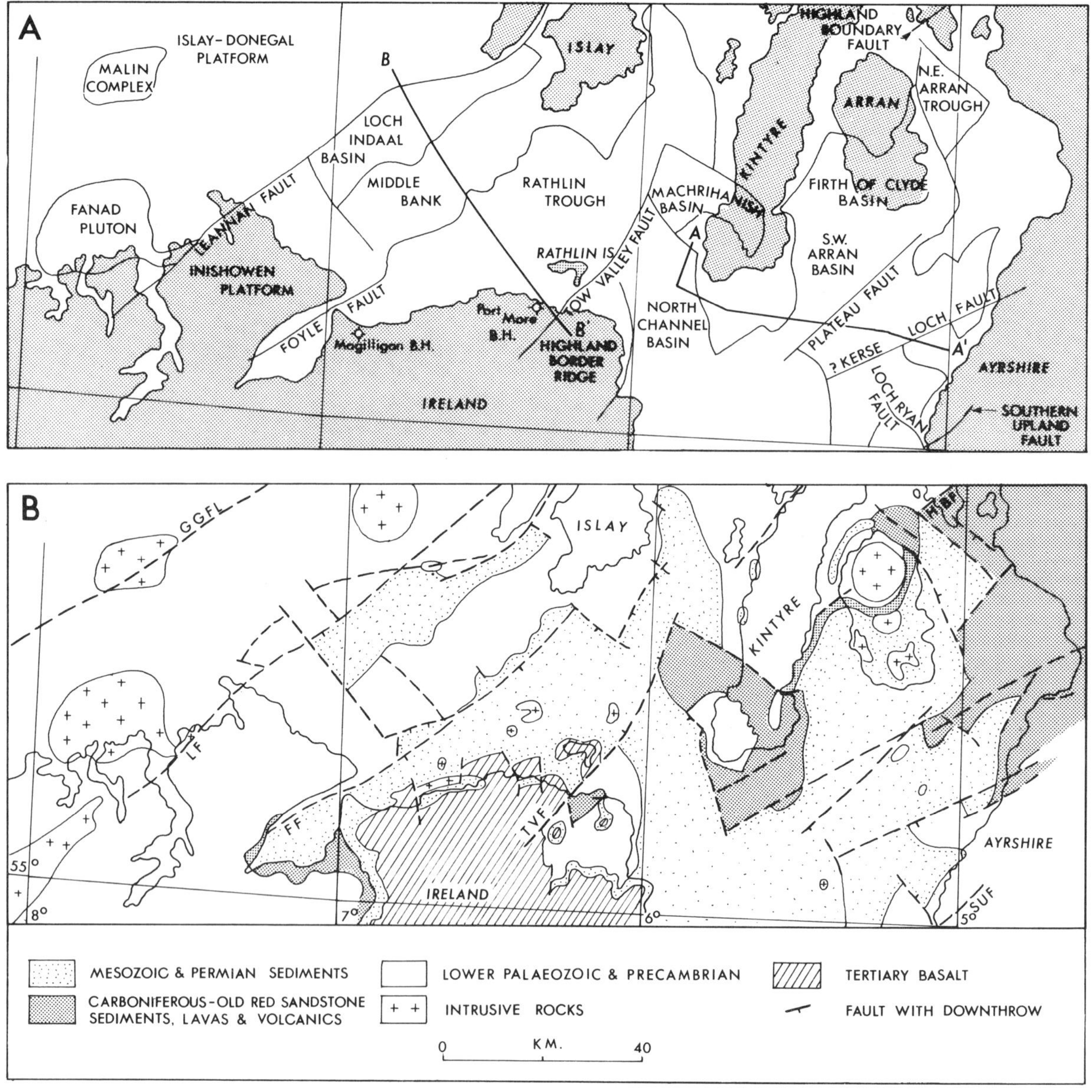

Figure 12.1 A. Structural elements in the region between southwest Scotland and northeast Ireland (in part after Evans *et al.* (1980).[7]) A–A′ and B–B′ denote the lines of section shown on Fig. 12.2.
B. Solid geology of the region between southwest Scotland and northeast Ireland (in the main derived from Evans *et al.* (1980)[7] and McLean and Deegan (1978).[14]) GGFL = Great Glen Fault Line, LF = Leannan Fault, FF = Foyle Fault, TVF = Tow Valley Fault, SUF = Southern Upland Fault, HBF = Highland Boundary Fault.

Although the principal development of the Midland Valley as a fault-bounded depression took place in Mid-to Late Devonian times, [2, 3] minor movements, associated with considerable volcanic activity, continued to take place throughout the Carboniferous. At the same time a shallow shelf sea invaded the depression, depositing marine and near-shore lagoonal sediments over both central Scotland and the northern part of Ireland. In Scotland, the Carboniferous was dominated by a continuous influx of erosional debris from the adjacent land areas of the Highlands and Southern Uplands, and in consequence is characterized by a sequence of sandstones, shales, thin limestones and coal seams. It is of note that some of the shale bands are rich in hydrocarbons and gave rise to a profitable oil shale industry during the nineteenth century and the early part of this century. To the southwest during the Carboniferous, a clear shallow shelf sea dominated the Magee Trough depositing a thick limestone sequence similar to that occurring in central and southern Ireland.

Following a period of continental conditions, the sea once again invaded the region during Upper Permian-Lower Triassic times, and extended eastwards from Rockall Trough to flood parts of the Firth of Clyde, the north of Ireland and the north Irish Sea. In other parts of the Firth of Clyde and the north Irish region continental conditions still persisted, with the deposition of thick sequences of desert sandstones and conglomerates. A return to widespread marine conditions took place in the Jurassic when the Liassic Sea extended westwards across the British Isles to link up with the Rockall Trough seaway. Fragmentary evidence suggests that much of northern Ireland, the Firth of Clyde and the Sea of Hebrides were originally overlain by a thin horizon of Jurassic marine sediments, most of which were rapidly worn away during the subsequent period of Lower Cretaceous uplift (Late Cimmerian phase). Lower Jurassic sediments are known from borehole evidence to exist as a thin layer of shales and limestones throughout the Ulster Basin. However, in the Firth of Clyde the only evidence for the northward extension of this Lower Jurassic horizon is represented by small fragments preserved within the volcanic complexes of Arran.[4] Elsewhere over the floor of the Firth, south of Arran, there is no evidence of any Jurassic or Cretaceous strata.

By the beginning of the Late Cretaceous, a shallow embayment of the Chalk Sea extended eastwards across the shelf from Rockall Trough submerging much of the Inner Hebrides, Firth of Clyde and Ulster region. A thin layer of basal sandstone overlain by chalk was laid down unconformably across the denuded surface of Jurassic, Triassic or older rocks. As with the Jurassic sediments, much of this chalk layer has subsequently been removed by erosion, and indications of its former extent are now only provided by a few isolated remnants throughout the region, such as the blocks preserved within the centre of an extinct volcano on Arran. However, in Ireland the Upper Cretaceous is still present in both the Rathlin and Magee Basins as a layer of hard white chalk, overlying a basal unit of greensand. Boreholes indicate that in places this succession may reach thicknesses of up to 150 m.

During the outburst of volcanic activity which accompanied the initial rupture of Greenland from Rockall during the Tertiary, large volcanic vents broke through to the surface in both northeast Ireland and the Firth of Clyde, and gave rise to outpourings of extensive plateau-like sheets of basaltic lava.

Geophysical data

There is relatively little deep seismic data available in this region and knowledge of the form and depth of the sedimentary basins has been gained from magnetic surveys,[5, 6] gravity models and the regular grid of sparker data.[7] In the case of the Loch Indaal Basin the gravity data suggest a sediment thickness of 2.4 km, whilst sparker traverses indicate a northwesterly regional dip into the Loch Gruinart Fault. The basin wedges out towards the northeast with unconformable overlapping of the Mesozoic sediments onto Dalradian basement rocks (Fig. 12.2A).

As in the case of the Loch Indaal Basin the Rathlin Trough is a region of low Bouger anomalies, with the lowest values in the area of Rathlin Island. Interpretation of the gravity and sparker data has been greatly aided by three deep reflection seismic lines acquired by the Institute of Geological Sciences.[7] A picture emerges (Fig. 12.2A) of a basin which deepens gradually with faulted downsteps eastwards towards the Tow Valley boundary fault and also southwards towards Rathlin. A well-defined seismic reflector is thought, by extrapolation to the onshore Port More borehole (Table 12.1 and Fig. 12.1A), to equate with the base of the Permo-Trias (which is here more than 2 km thick). The layered interval seen on the seismic sections beneath the reflector would probably then represent Carboniferous strata. Shallow sparker data and gravity modelling suggest a partly faulted and partly gradational northern margin to the Rathlin Trough. The geophysical data also indicate the presence of a number of igneous bodies intruded into the Mesozoic sequence, particularly in the area of Rathlin Island. Some of these intrusions form seabed topographic features.

The North Channel and the Firth of Clyde have been the focus of several geophysical research programmes, initiated by the universities but with considerable later involvement by the Institute of Geological Sciences.[7] During the 1960s seaborne gravity together with

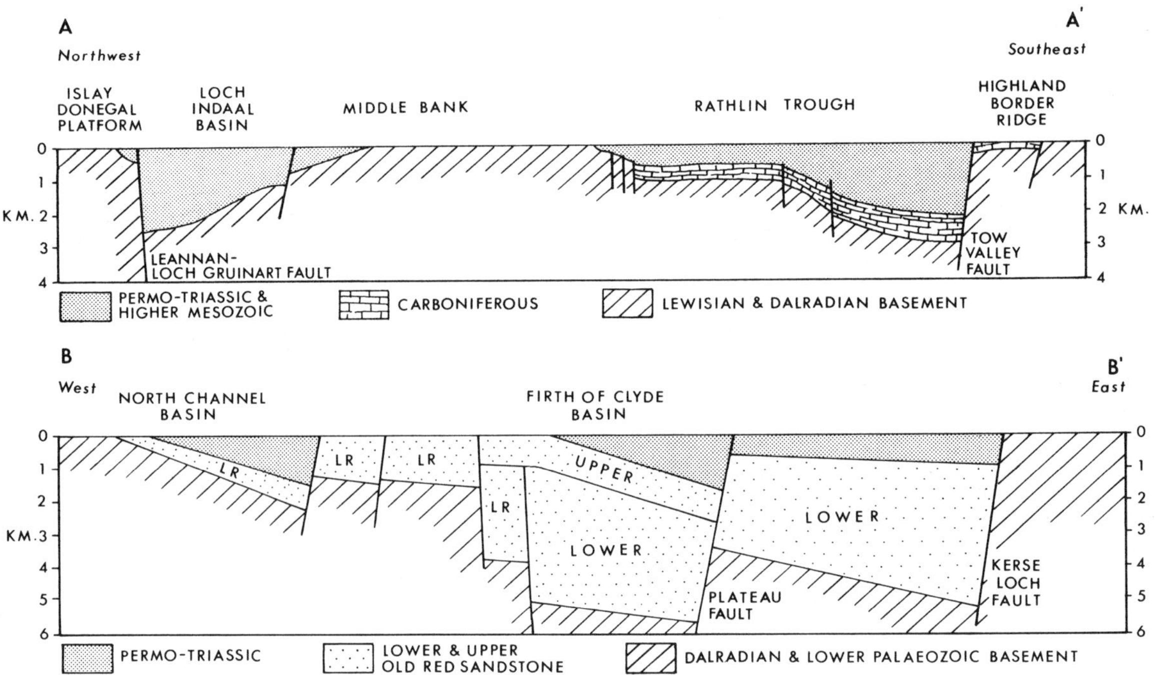

Figure 12.2A–A′ Diagrammatic cross-section of the Loch Indaal Basin and Rathlin Trough (adapted from Evans *et al.* (1980)[7]).
B–B′ Diagrammatic cross-section between the Mull of Kintyre and the Ayrshire coast (in part adapted from McLean and Deegan (1978)[14]).

aeromagnetic and marine magnetic surveys and seismic refraction work established the structural framework and basin form in the region. Later work involved detailed sparker, boomer and pinger surveys together with a limited amount of commissioned reflection seismic traversing. The geophysical programmes have been followed by intensive seabed sampling and drilling.

There is a considerable degree of agreement between the different geophysical surveys of these southern basins. The northeast-southwest structural regime is again dominant. However there is some doubt concerning the line of the Highland Boundary Fault. This may be traced on the Scottish mainland to the Firth of Clyde and then haltingly southwestwards across northwest-striking faults to the igneous centre on the Arran. Its course beyond Arran is more conjectural and although it may be projected as far as Kintyre there must be doubt on geophysical grounds that there is a connection across the North Channel with the Cushendall Fault in Ireland at the southern margin of the 'Highland Boundary Ridge' sometimes quoted as the Highland Boundary Fault extension.[11] Indeed the Cushendall Fault may link with the Plateau-Dusk Water Fault trend in Scotland,[8] which similarly throws Carboniferous rocks down to the northwest, in contrast to the normal Highland Boundary Fault configuration.

The Southern Upland Fault is traced without difficult to Loch Ryan on the west coast of Scotland at 55° North latitude (Fig. 12.1) where it is bounded by Lower Palaeozoic rocks on both sides. A number of subsidiary New Red Sandstone basins are developed between the Tow Valley Fault and the Southern Uplands Fault, and controlled by the intervening major faults. The North Channel Basin is a complex structure containing Upper Palaeozoic rocks and Permo-Triassic rocks (Fig. 12.2B). An extensive band of New Red Sandstone extends southwards from Jura to link with outcrop along the coast of Antrim. Seismic profiling reveals a number of northwest-striking faults crossing the main northeast structures.

In what might broadly be termed the *Firth of Clyde Basin* Permo-Triassic rocks encircle Arran and thicken southwards into the *Southwest Arran Trough* and thence extend to the Plateau Fault (Fig. 12.1). The geophysical evidence indicates, as in much of this region, that pre-Permo-Triassic depocentres existed in Upper Palaeozoic time. Interpretation of gravity and magnetic data is made difficult by the coincidence of Upper Palaeozoic and younger basins, particularly where New Red Sandstone redbeds rest on thick Old Red Sandstone sediments. Geophysical evidence indicates that the spread of New Red Sandstone south of the Plateau Fault represents a thin cover resting directly on Lower Old Red Sandstone rocks (Fig. 12.2B).

In Scotland there are a number of fault bounded basins – the Mauchline and Sanqhar Basins – which extend across the Southern Uplands element. A similar structure is present at the coast in the *Stranraer Basin* which extends southwards from the Southern Uplands Fault line, bounded on the east by the Loch Ryan Fault (Fig. 12.1A). The basin fill is comprised of some 2.0 km of New Red Sandstone underlain by thin Upper Carboniferous strata. The latter consist of red and grey sandstones and shales with thin basalts.

Stratigraphy

The Rathlin Trough and the Midland Valley were already in existence as major depressions by the end of Devonian time. In consequence there are extensive developments of Old Red Sandstone and Carboniferous rocks beneath, and at the margins of, the Mesozoic basins.

The *Loch Indaal Basin* may contain as much as 2.4 km of Mesozoic strata, being fault-bounded in the northwest and entirely rimmed by Lewisian and Dalradian rocks. Structure within the basin appears from sparker evidence to be relatively simple, with low northwesterly dips towards the thicker parts of the basin against the Leannan-Loch Gruinart Fault. Two types of sedimentary rocks, both with distinctive seismic signature, have been recognized within the basin; sediments with a strong reflecting character generally in the northwest, and seismically homogeneous (and probably older) rocks in the southeast. Boreholes[7] have demonstrated the older unit to consist of gypsiferous marls presumed to belong to the Keuper whilst the younger rocks (marls and mudstones) are of Rhaetic, Liassic and possibly younger age. It is likely however that the greater thickness of basin fill consists of Triassic redbeds.

The *Rathlin Trough* lies to the northwest of the Highland Border Ridge. Mesozoic rocks crop out along the north coast of Ireland,[13] but are mainly covered by plateau basalts inland. Towards the western margin of the trough the Magilligan borehole penetrated a Lower Lias to Permian sequence to a depth of 976.3 m resting on Carboniferous rocks with thin coals. Further east a borehole at Port More was in probable Permian sandstones when terminated at 1864 m (Table 12.1). It seems reasonable to assume that the onshore basin contains in the order of 2,500 m of Mesozoic strata (Figs. 12.2A and 12.3). This is borne out offshore where a strong seismic reflector correlates with a level of about 2,000 m, just below the base of the Port More borehole, and an estimated further 600 m can be seen below the reflector offshore. It has been suggested[7] that the borehole bottomed in Permian rather than Bunter sediments, in which case the offshore reflector might approximate the base Permo-Triassic, with the underlying strata being of Carboni-

TABLE 12.1 Rathlin Trough

Magilligan Borehole	*Spudded June 1963*	*Completed January 1964* *Thickness (m)*	*Base (m)*
Recent	Drift	52.4	52.4
Tertiary	Sill	58.6	111.0
L. Jurassic (L. Lias)		89.9	200.9
Rhaetic		7.9	208.8
Keuper	Marl	371.5	580.3
Bunter/Permian	Sandstone	396.0	976.3
Carboniferous		370.3 T.D.	1346.6
Port More Borehole	*Spudded March 1965*	*Completed February 1967* *Thickness (m)*	*Base (m)*
Tertiary	Basalt	77.1	77.1
Cretaceous	Upper Chalk	91.3	168.4
L. Jurassic (L. Lias)	Mudstones	269.4	437.8
Tertiary	Igneous sill	222.9	660.7
U. Triassic (Rhaetic)	Shale	4.1	664.8
Triassic (Keuper)	Marl-siltstone	652.5	1317.3
Triassic (Bunter)	Sandstone	51.3	1797.7
Permian	Sandstone	66.3 T.D.	1864.0 m

ferous age. The general picture which emerges from the geophysical data is of a basin which deepens southeastwards towards the Tow Valley Fault. The structure of the basin is simple except for zones of tight folding in the vicinity of major faults.

The plateau basalts, extensive onshore, extend only a short distance offshore, notably around Rathlin Island. However a number of basic intrusive bodies have been identified, scattered across the central portion of the offshore basin. Where these crop out on the seabed they produce topographic mounds. Dykes and sills occur extensively, but patchily, in the sediments of the Rathlin Trough. A small number of offshore boreholes have demonstrated a Liassic grey mudstone sequence overlying redbeds of presumed Permo-Triassic age with the succession younging southeastwards.

East of the Tow Valley Fault a marine borehole and shallow seismic profiles indicate an offshore extension of Carboniferous rocks at Machrihanish on Kintyre (Fig. 12.1A), bounded in the north by the Kilchenzie Fault and in the west by the Tow Valley Fault extension. The Mesozoic rocks of the northward extension of the North Channel Basin breach the Highland Boundary Ridge and extend to the Tow Valley Fault. The thickness of Permo-Triassic rocks hereabouts may be in the order of one kilometre, with perhaps thin underlying Carboniferous sediments. Further south towards Larne the Mesozoic fill of the North Channel Basin may be considerably thicker (Fig. 12.3).

In the *Magee Basin* Bunter facies sandstones overstep northwards onto the longstanding positive element of the Highland Boundary Ridge, and thin to 200 m. At Murlough Bay, on the north side of the ridge, 26 m of strata represent the Triassic. Permian sandstones, which were probably present in the Port More borehole, do not appear to be represented in the Magee Basin.

The onshore succession of the Magee Basin has been penetrated in a number of deep boreholes (Fig. 12.3). At Longford Lodge a borehole penetrated a Chalk, Lias, Rhaetic and New Red Sandstone sequence beneath basalts, finally entering metamorphic basement at 1,452.7 m. The Triassic redbeds are here 617.5 m thick, with sandstones grading up to marls. The Larne-l borehole (Fig. 12.1A and Table 12.2) bottomed in Permo-Triassic sandstones at 1,283.5 m after penetrating almost 1,000 m of 'Keuper' marls containing three major and several minor salt horizons (totalling 533 m) and nine basalt intrusions. Twenty metres of dark shales were ascribed to the Rhaetic. The Lias is generally thin in the basin and generally follows conformably on the Rhaetic, reflecting the distribution and thickness patterns of the underlying sequence.

The Triassic succession in the offshore North Channel Basin probably reaches 2 km or more in thickness, thinning northwards against the Highland Border Ridge, and also by overlap towards the eastern basin margin (Fig. 12.3). A succession similar to that of the Larne borehole might be anticipated, although halokinetic structures have not been observed on the geophysical records.

The onshore succession continues with Rhaetic and Lower Liassic beds. The remainder of the Jurassic and the Lower Cretaceous are not represented, although Middle and Upper Liassic fossils are found in the basal beds of the overlying unconformity at Murlough Bay on the flanks of the Highland Border Ridge, and in the drift deposits around Ballycastle. Basal Greensands of Upper Cretaceous age (Cenomanian-Turonian) are followed by widespread Chalk (to 110 m) of Late Senonian to Early Maastrichtian age. This unit probably originally covered the whole area, including the Highland Border Ridge, before being reduced by erosion. The Chalk is more indurated than its counterpart in the Upper Chalk of England, a fact which has been attributed to

TABLE 12.2 Larne Borehole. Magee Basin

Spudded August 1962		*Completed April 1963*	
Succession	*Lithology*	*Thickness (m)*	*Base (m)*
Glacial drift	Boulder clay	36.6	36.6
Lias	Mudstones	51.5	88.1
Rhaetic	Siltstone-mudstone	20.5	108.6
Keuper	Argillites	254.4	363.0
Keuper Larne Halite	Salt	481.2	844.2
Keuper	Halite & Argillite	231.2	1075.4
Bunter	Sandstone	208.1 T.D.	1283.5

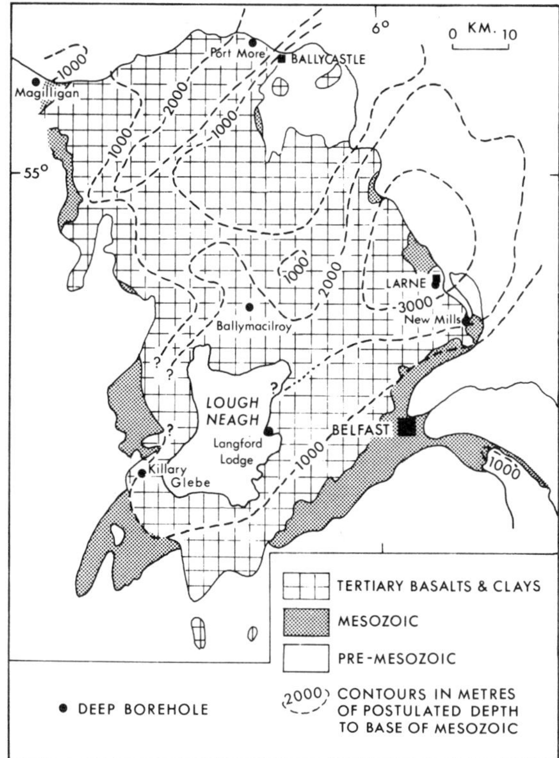

Figure 12.3 Onshore Mesozoic sedimentary basins, northeast Ireland (modified after Thompson (1979)[11]).

hydrothermal metamorphism resulting from burial under a thick pile of Tertiary basalts. The sequence of depositional events in the region was terminated by the Eocene Plateau basalts which are often in excess of 400 m thick. In addition the region is crossed by a profusion of intrusive dykes, with related sills and small stocks. The onshore Rhaetic to Tertiary succession is barely represented in the offshore and certainly there is no thick development of marine Mesozoic strata in the North Channel and Firth of Clyde offshore basins.

In the *Firth of Clyde Basin* there is a thin rim of New Red Sandstone on the west and east of Arran but the main basins lie on the northeast and southwest of the island (Fig. 12.1A). The *Northeast Arran Trough* is shown by geophysical data to contain a thick sequence (1.5 km) of Upper Palaeozoic and New Red Sandstone sediments. The basin is virtually a fault-bounded triangle, with the Bute Sound and Brodick Bay Fault to the east and west, respectively, and the Dusk Water Fault extension in the south. The basin fill thickens southwards towards the Dusk Water Fault and post-Triassic sediments have not been proven.

The sediments in the *Southwest Arran Trough* are a thicker extension of the sequence known onshore in southern Arran. The thickest part of the basin is immediately northwest of the southwest-trending southern boundary fault, the Plateau Fault, with perhaps 3.5 km of Upper Palaeozoic (mainly Old Red Sandstone) and 1.0 km of New Red Sandstone. Boreholes within the basin indicate rocks of Keuper facies at surface and no younger strata are known. The basin links in the southwest into the North Channel Basin. Southeast of the Plateau Fault, the Plateau area has a thin cover of New Red Sandstone facies resting on Lower Old Red Sandstone rocks.

From this brief review of the Ulster Basin it will be seen that Tertiary sediments are not found in the offshore areas. There is of course a thick development of Tertiary plateau lavas in Antrim but elsewhere Tertiary sedimentary rocks are only poorly represented. The exception to this is in the Lough Neagh Basin further south where a sequence of Oligocene-early Miocene fluvial and lacustrine sediments totalling over 350 m were deposited in a local structural depression.

Petroleum Potential

Thick Upper Palaeozoic and Permo-Triassic sequences are found in the area within the southwest extension of the Midland Valley of Scotland. Onshore drilling indicates that both the Bunter and Keuper successions contain prospective reservoir sands, and Permian sandstones also occur over part of the basin. The intervening basement highs, even where covered by a thin veneer of younger rocks, have a low exploration potential. The thick evaporites penetrated by onshore drilling in the Magee Basin add further promise to the area. Fault-controlled structures and seal may thus exist in the basins but a question must be placed against source potential.

The Carboniferous rocks of the Antrim coastal belt are Viséan-Namurian limestones, sandstones and shales with coals. At Machrihanish in southwest Kintyre there is a Viséan to Westphalian sequence comparable to that at Ballycastle. Carboniferous rocks are thought to extend offshore into the Rathlin Trough (Fig. 12.2A). However seismic data between the Mull of Kintyre and the Irish coast[8] indicate that the Carboniferous is absent. Further north there may be a thin coal-bearing sequence from the west of Kintyre to Ballycastle in Antrim. There is thus a possibility for gas generation from the coal-bearing strata over parts of the offshore area, and areas with Keuper salt seal could be prospective. Doubts must exist concerning the potential and maturity of any source rocks in the Mesozoic sequence. Over much of the offshore area there is little evidence to suggest depths of burial significantly greater than those existing at the present time.

References

1. DOBSON, M. R. and EVANS, D. 1974. The geological structure of the Malin Sea. *J. geol. Soc. London* **130**, 475–478.
2. MACGREGOR, M. and MACGREGOR, A. G. 1948. The Midland Valley of Scotland. *British Regional Geology*. H.M.S.O. 95 pp.
3. GEORGE, T. N. 1960. The stratigraphical evolution of the Midland Valley. *Trans. Geol. Soc. Glasgow* **24**, 32–107.
4. RICHEY, J. E. 1948. Scotland: The Tertiary Volcanic Districts. *British Regional Geology*. H.M.S.O. 105 pp.
5. RIDDIHOUGH, R. P. 1968. Magnetic Survey off the north coast of Ireland. *Proc. R. Irish Acad.* **66B**, 27–41.
6. RIDDIHOUGH, R. P. and YOUNG, D. D. G. 1971. Gravity and magnetic surveys off the north coast of Ireland. *Proc. geol. Soc. London* 1664, 215–220.
7. EVANS, D., KENOLTY, N., DOBSON, M. R. and WHITTINGTON, R. J. 1980. The geology of the Malin Sea. *Rep. Inst. geol. Sci. London* 79/15, 44 pp.
8. FRIEND, P. F., HARLAND, W. B. and HUDSON, J. D. 1963. The Old Red Sandstone and the Highland Boundary in Arran. *Trans. geol. Soc. Edinburgh* **19**, 363–425.
9. McLEAN, A. C. and DEEGAN, C. E. 1978 (*Eds.*). The solid geology of the Clyde Sheet (55°N/6°W). *Rep. Inst. geol. Sci. London* 78/9, 114 pp.
10. FRIEND, P. F. and MacDONALD, R. 1968. Volcanic sediments, stratigraphy and tectonic background of the Old Red Sandstone of Kintyre, W. Scotland. *Scott. J. Geol.* **4**, 265–282.
11. WILSON, H. E. 1972. Regional geology of Northern Ireland. H.M.S.O. 113 pp.
12. SIMPSON, J. B. and RICHEY, J. E. 1936. The geology of the Sanquahar Coalfield and the adjacent basin of Thornhill. *Mem. geol. Surv. G.B.*
13. WILSON, H. E. and MANNING, P. I. 1978. Geology of the Causeway Coast. Vol. 1, *Mem. geol. Surv. N. Ireland*. H.M.S.O. Belfast, 72 pp.
14. THOMPSON, S. J. 1979. Preliminary Report on the Ballymacilroy No.1 borehole, Ahoghill, Co. Antrim. *Geol. Surv. N. Ireland Open File Rep.* 63, 15 pp.

Chapter 13

Irish Sea Basins

Introduction

The Irish Sea between the Southern Uplands and Longford Down Massifs in the north and Welsh Massif in the south concels three sedimentary basins. These are the *Solway Firth Basin*, the *Irish Sea Basin* (sometimes called the *Manx-Furness Basin*) and the *Kish Bank Basin*. The largest of these is the Irish Sea Basin which extends southwestwards onshore to link with the elongate *Cheshire Basin* (Fig. 13.1).

Gravity data show the Solway Firth, which lies immediately south of the Kirkudbrightshire coast in the northeast Irish Sea, to be underlain by a northeast-trending basin. This narrow trough, strongly influenced by the caledonoid trend of the basement rocks, extends from the northern tip of the Isle of Man towards the onshore Permo-Triassic basin of the Carlisle-Vale of Eden area. Permo-Triassic rocks representing an extension of the *Solway Firth Basin* are also exposed on the Isle of Man, and the sediments thicken offshore towards the centre of the basin.

The *Irish Sea Basin* is a much broader feature by comparison with the other basins of this region and is less obviously influenced by the caledonoid structural framework. Many of the intrabasinal faults have a northwest-southeast (Charnian) trend. The basin is partially separated from the Solway Firth Basin by the narrow *Ramsey-Whitehaven Ridge* in the north. Southeastwards it extends from the Isle of Man towards the Lancashire coast where it is contiguous with the coastal Permo-Triassic rocks of the Liverpool-Wirral area and thence is connected to the deep, fault-bounded Cheshire Basin onshore. Water depths over the basin are less than 50 m.

The Irish Sea Basin has also been the focus of oil company exploration. Seismic reflection surveys have been followed by drilling, with consequent greatly improved understanding of the stratigraphy. The discovery of a commercial gas field (the Morecambe Gas Field) within this basin has led to great exploration interest.

The existence of the *Kish Bank Basin* was first indicated by geophysical surveys in the early 1960s. It lies only a few miles east of Dublin in relatively shallow water of the order of 30–60 m deep. In places it is covered by as little as 20 m of water. The basin is largely covered by the four offshore blocks (33/16, 33/17, 33/21 and 33/22) which were licenced to Amoco in 1976, although it is thought to extend somewhat further east and slightly south of these blocks (Fig. 13.1). However, following the completion of two dry holes, one by Amoco in 1977 and the second on a farm-in basis by Shell in 1979, the licence was relinquished in 1979. The blocks have been included among those on offer by the Irish Government in the Second Licencing Round of late 1981-early 1982. The Kish Bank Basin differs from the other two basins of the Irish Sea in that it is entirely an offshore feature and no post-Palaeozoic rocks are found on the adjacent coastline.

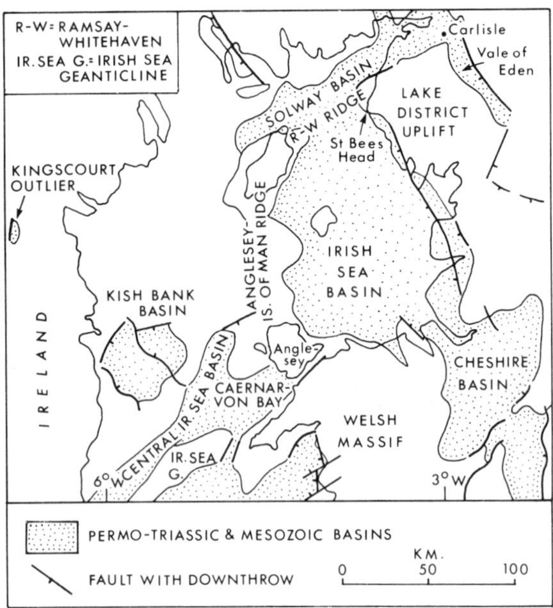

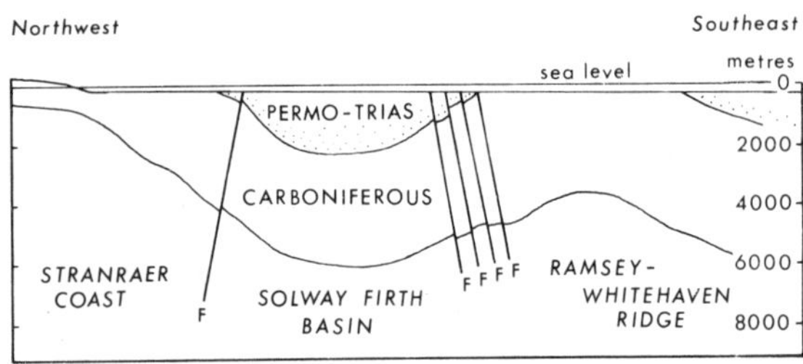

Figure 13.1A. Schematic relationships of Permo-Triassic Basins around the Irish Sea.
B. Diagrammatic cross-section of the Solway Firth Basin.

Geophysical Data

The first indications that the Irish Sea was underlain by a number of very thick young sedimentary basins (which are either absent or in general poorly exposed onshore around the margin of the sea) became apparent after regional gravity surveys had been completed across the area.[1, 2] These surveys, carried out by a number of universities and by the Institute of Geological Sciences, showed the Irish Sea to be dominated by generally high gravity values in comparison with the surrounding land areas, superimposed across which are a series of local anomalies. These anomalies correspond to the shape of the underlying basement structure, with the regions of relatively low gravity values correlating with the areas underlaid by sedimentary basins, and regions of relatively high gravity values correlating with the uplifted basement features.

Marine magnetic and aeromagnetic work over the central and northern parts of the Irish Sea shows a broad pattern of alternating belts of positive and negative anomalies which outline the main structural and basinal elements. During the period 1967–1969, the north Irish Sea was covered by a programme of shallow seismic reflection profiles,[3] giving variable depths of penetration beneath the sea bed of either 100 m or 300 m, although in areas of greatest interest penetration was increased to 500 m. These profiles show an upper seismic layer less than 40 m thick, overlying most of the area, and a lower composite layer consisting of areas of acoustic and economic basement flanked by regions of structurally more complex nature showing a number of internal reflectors and evidence of flat-lying or folded strata. For the purpose of interpretation, the velocity of this upper layer was initially assumed to be 1.8 km/sec and that of the underlying layer as 3.0 km/sec. However, subsequent refraction work in 1971 indicated a velocity of 3.8 km/sec for sediments beneath the upper seismic layer which infilled a sedimentary basin, with higher velocities at deeper levels outside the basin.

Over the northeastern part of the Irish Sea, shallow seismic reflection data reveal the presence of flat-lying and gently folded beds within the *Irish Sea Basin*, which appear to be a continuation of the Permo-Triassic coastal outcrops of Lancashire and Cumberland. Shallow coring of the sea bed by the Institute of Geological Sciences has established the presence of red Triassic mudstones, while onshore boreholes close to the Lancashire coast have encountered several thousand feet of Permo-Triassic rocks.

Further north, shallow seismic reflection data have also confirmed the presence of a sediment-filled trough underlying the Solway Firth (see Fig. 13.1B). Magnetic and seismic data suggest that the margins of this basin may be fault controlled. The seismic profiles indicate a two-fold sequence of sediments infilling the *Solway Firth Basin*, comprising a thick lower group characterized by poor seismic reflectors, which may represent Permo-Triassic strata, and a thin upper group, restricted to the central axis of the basin, of probable Lower Jurassic rocks. Following the northern margin of the basin is a narrow belt of structural disturbances which is believed to indicate the presence of salt within the basin.

Gravity work in the area immediately east

of Dublin carried out by academic institutions (notably the University of Durham) in the mid to late 1960s revealed the presence of a fairly major negative anomaly in the area of the Kish Bank. This was interpreted as reflecting a sedimentary basin containing in the order of 3 to 4 km of early Mesozoic fill.[1] Speculative seismic reflection lines shot during the period 1970 to 1975 confirmed the presence of such a deep fault-controlled basin. Further detailed seismic surveys carried out by Amoco in 1976 to 1978 gave another 1,176 km of reflection seismic data, in addition to gravity and magnetic profiles and completed the geophysical exploration of the basin to date. This large amount of geophysical information has allowed the fairly exact delineation of the extent and nature of most of the Kish Bank Basin.

Stratigraphy

Solway Firth Basin

No deep drilling has taken place in the Solway Firth Basin but the basin extends onshore into the Carlisle and Vale of Eden Basins (Fig. 13.1) which afford a view of the stratigraphy which is pertinent to an understanding of the sequence to be expected offshore.[4–7]

The lowest sediments (Fig. 13.4) rest directly on the eroded surface of Palaeozoic rocks, and are composed of coarse sandstones and pebble beds of Permian age derived by rapid erosion of the surrounding uplands. These are overlain by a thick (up to 300 m) red dune sandstone, the Penrith Sandstone. The presence of reptilian footprints preserved in horizons within the sandstone unit, suggests the periodic existence of lakes and pools during this continental, semi-arid period of deposition. Much finer grained rocks, consisting mainly of dull red mudstones (the Eden Shales and Evaporites; 120 m) overlie the Penrith Sandstone, and are themselves overlain by a further succession of sandstones (St. Bees Sandstone 500 m ± and Kirklinton Sandstone, up to 125 m). The overlying red mudstone sequence (Stanwix Shales; up to 300 m, ? Middle-Upper Triassic) lacks the thick evaporites commonly found in Keuper. Although much of this Permo-Triassic sequence appears to have been deposited under terrestrial, desert-like conditions, there is some evidence of marine influence. At several levels, beds of gypsum and anhydrite, and occasionally thin dolomites containing marine fossils, interfinger with the dune sandstones and shales, suggesting that the onshore part of the Solway Firth Basin may have suffered infrequent flooding by the sea from the west.

It is probable that the Eden Shales, on the basis of the contained flora, are Late Permian in age. The overlying thick sandstone sequences are unfossiliferous but probably straddle the Permian-Triassic boundary. There is no direct palaeontological control and the Permian-Triassic boundary has variously been placed within the St. Bees Sandstones,[5] or at the top or within the St. Bees Evaporites.[7] Low gravity values over the Vale of Eden suggest that more than 1 km of Permo-Triassic rocks are present northeast of Penrith.

West of Carlisle is a poorly exposed area of Lower Jurassic (Liassic) rocks comprising a sequence of dark shales with thin limestone bands. These suggest deposition under marine conditions at a time when the Jurassic seas made large incursions across the old Permo-Triassic landscape. In summary, it appears that the Carlisle-Vale of Eden area is the landward extension of the present-day Solway Firth Basin, as it was during Permo-Triassic times. It is probable therefore, that the offshore area contins thicker sequences of marine rocks than are evident onshore.

Cheshire Basin

Although this book is concerned chiefly with the offshore areas west of Britain, and there are major differences between the overall geology of the land and sea areas, the coastline often forms a very imprecise boundary between the two provinces. Nowhere is this better demonstrated than in the landward extension of the Irish Sea Basin into the *Cheshire Basin and Worcester Graben*. For this reason, a brief account of the onshore basins is included here.

The Cheshire Basin (Fig. 13.2) is the low-lying area of Permian and Triassic rocks, extending from the Pennine uplands in the east to the Welsh Border hills in the west. South of Kidderminster, the fault-controlled basin narrows to a tongue extending through Worcester towards Gloucester.

Rocks of Upper Carboniferous age (including the Coal Measures) form a broken and discontinuous ring around the Cheshire Basin. Coal Measure rocks originally formed a continuous sheet across central England as far west as the Welsh Mountains. It is probable that they still floor much of the Cheshire Basin although ribs of older rocks may, in places, protrude up into the Permo-Triassic strata. A well at Knutsford in the northern part of the Cheshire Basin penetrated Westphalian sandstones, shales and coals whilst Prees-1 further south (Figs. 13.2 and 13.3) intersected 161 m of Upper Westphalian reddened shales before bottoming in indurated Lower Palaeozoic rocks. The Hercynian earth movements which occurred at the end of the Carboniferous period, uplifted and folded the strata and resulted in an extension of the Welsh Mountain massif and produced the upland spine of the Pennines.

The outline stratigraphy of the Cheshire Basin was known for many years from outcrop and shallow borehole data, but our knowledge has been considerably enhanced in recent years by deep drilling. Early oil exploration in the Cheshire Basin was concentrated in the Formby area on the coast in the north (Fig. 13.2), where oil seeps were known. Deep drilling over more than two decades resulted in a number of holes which penetrated through the Permo-Triassic sequence into Carboniferous rocks.[10] The two wells, B.P./British Gas Knutsford-1 and Trend Exploration Prees-1, drilled in the early 1970s, added greatly to an understanding of the central part of the basin (Figs. 13.2 and 13.3). For ease of discussion the traditional names for the lithostratigraphic subdivisions are used here. However, for more detailed subdivisions and description together with a comprehensive stratigraphic framework the reader is referred to the work of Thompson,[11] and of Warrington *et al.*[7]

The Permo-Triassic rocks were primarily deposited in an arid, terrestrial environment. Faults define the present basin outlines and probably controlled the development of the basins during accumulation of the sediments. The products of erosion from the Hercynian uplands collected within these fault-bounded intermontane basins, resting unconformably on the underlying surface of erosion.

Within the Cheshire Basin, Permian rocks are recognized at the surface with certainty only in the Manchester area, where a thick basal sequence of sandstones is overlain by a tongue of marine sediments, the Manchester Marls, containing a fauna of Permian (Zechstein) aspect. It may be presumed that this temporary marine invasion proceeded from the north along the eastern margin of the basin, although its precise extent beneath the basin is unknown. Elsewhere in the basin, the rocks form an unbroken sequence of dominantly red non-marine rocks, some of which may be equivalent in age to the Manchester Marl. However, it is impossible to draw a precise boundary between Permian and Triassic strata and the whole sequence has long been known as the New Red Sandstone. Nevertheless, it is probable that rocks deposited during the Permian Period do, in fact, extend across much of the Cheshire Basin.

A simplified Permo-Triassic succession in this area can be regarded as follows:

Triassic	Rhaetic	
	Keuper	– red mudstones (marls with salt beds and sandstones in the lower part.
	Bunter	– sandstones and pebble beds.
	Manchester Marl	– marls, shales and calcareous sandstones.
Permian	Collyhurst Sandstone and basal breccias.	

It is not surprising that, during erosion of the landscape of central England following upon the Hercynian earth movements, breccias and pebble beds were laid down in parts of the central England basins. The boundary faults probably continued to move during deposition and thus breccia fans were concentrated along the fault scarps. At the northern margin of the Cheshire Basin, the Collyhurst Sandstone contains breccias and pebbles at the base and the unit thickens and thins across fault lines, suggesting contemporaneous control of deposition. The sandstones, which are up to 720 m thick in the Formby boreholes at the coast, are overlain by the Manchester Marl, a sequence of marls with thin calcareous beds. In the deep Prees-1 well in the south of the basin (Figs 13.2, 13.3) the Manchester Marl has been replaced by arenaceous equivalents, giving support to the idea of an incursion from the north. The Collyhurst Sandstone of the same well is indistinguishable from the Lower Bunter sandstones, thus giving an unbroken sequence of sandstones ranging up into the Triassic (Fig. 13.3).

The Triassic rocks were deposited in more uniform layers over a wider area of the Cheshire Basin than those of the Permian, and probably extended further southwards along the narrow Worcester Graben. The topographic relief in Triassic times had been gradually subdued by erosion and the Hercynian uplands reduced to a low peneplain. The terms Bunter and Keuper are used to describe the sand-dominant and marl/mudrock-dominant parts of the Upper Permian and Triassic without implying precise correlation with units of the same name in the North Sea Basins.

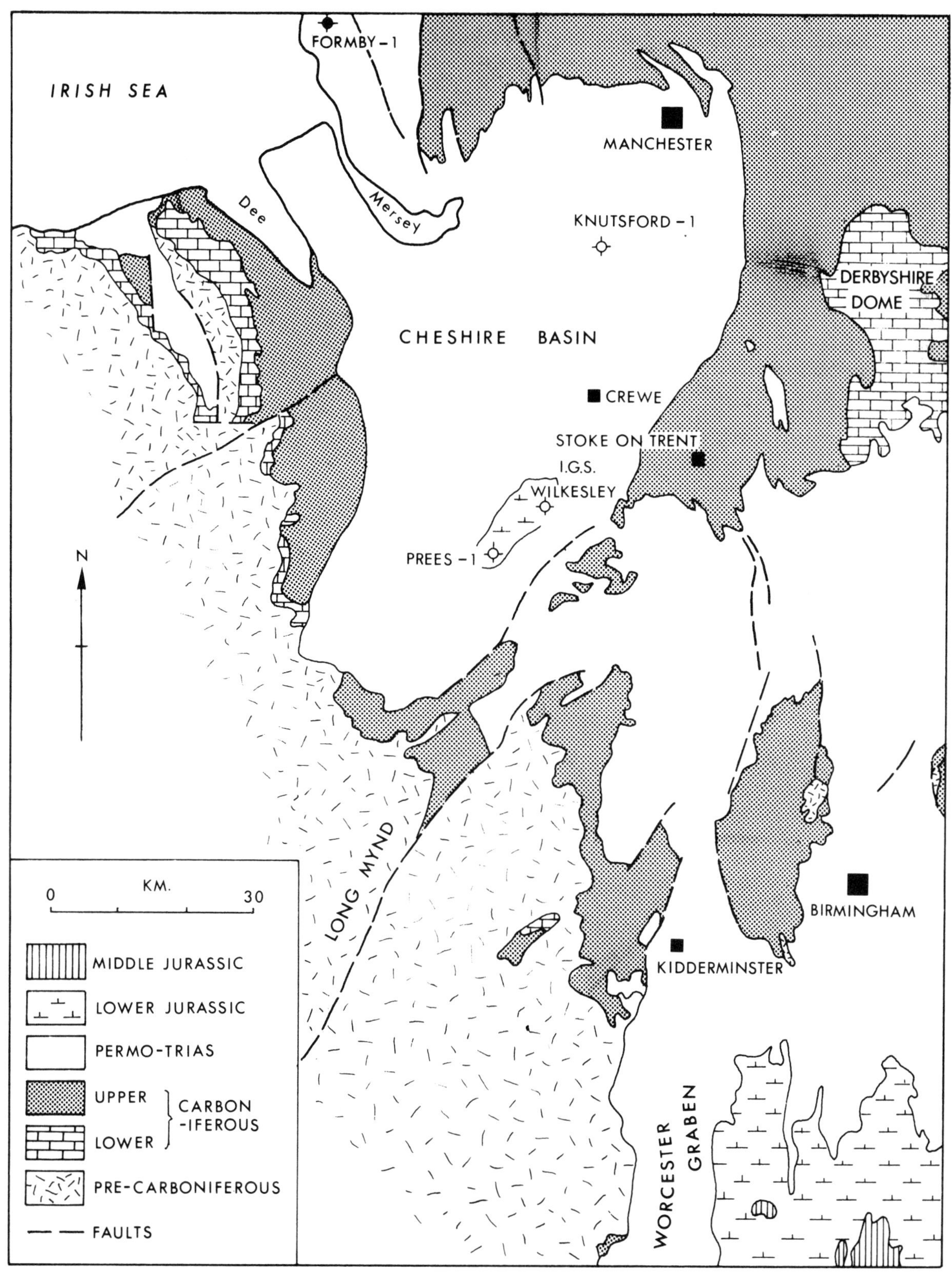

Figure 13.2 Geological sketch map of the Cheshire Basin.

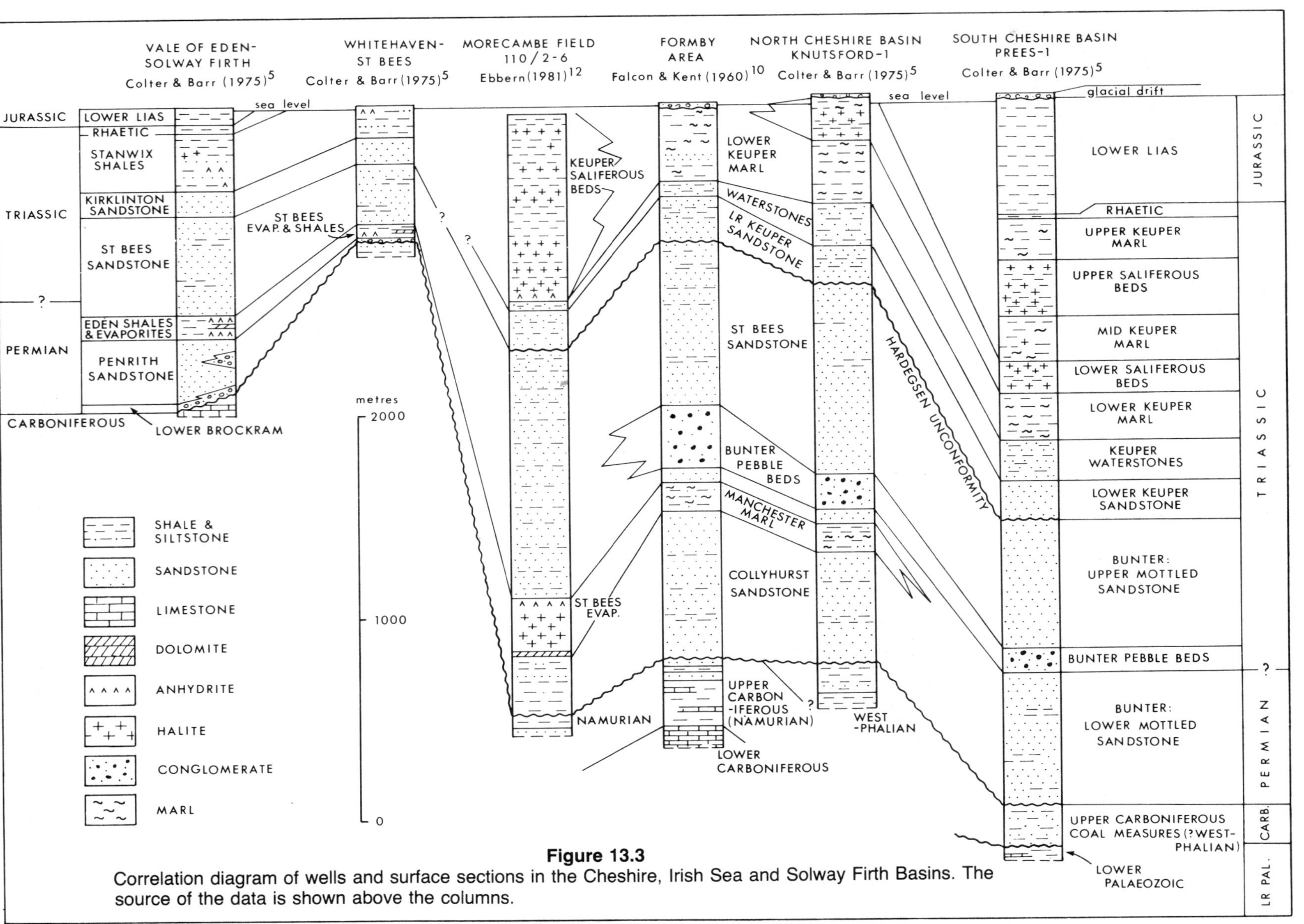

Figure 13.3
Correlation diagram of wells and surface sections in the Cheshire, Irish Sea and Solway Firth Basins. The source of the data is shown above the columns.

The thick Bunter sequence comprises predominantly red sandstones, many of which show the rounded, polished sand grains and other features typical of desert dune deposits. However, the Bunter pebble beds and many of the upper sandstones also exhibit the characteristics of water deposition. The pebbles were deposited as fans and river valley deposits resulting from northward flowing rivers originating in the Welsh Mountains. Some of the more unusual pebbles suggest transport along a river system flowing northwards from Cornwall or Brittany. Certainly the pebbles decrease in number and abundance northwards across Cheshire and Lancashire towards the offshore basin.

The Keuper deposits extend beyond the limits of Bunter deposition and, in places, rest unconformably upon folded pre-Permian rocks. In general, the lower part of the Keuper sequence is comprised of grey and mottled red sandstones.[11] These are regarded as being unconformable on the Bunter, the break in sedimentation being equivalent to the Hardegsen Unconformity on the North Sea area. Dessication cracks, reptile footprints and other features all suggest a continuation of arid conditions with intermittent rivers and shallow lakes. The flaggy sequence at the top of the sandstones and of the overlying Keuper Waterstones may represent lagonal deposition with weak marine influence, possibly equating with the Muschelkalk transgression of the North Sea Basin.

The Keuper Marl is a uniform and widespread unit which attains a thickness of several thousands of feet in the centre of the Cheshire Basin. The unit is composed for the most part of red and chocolate coloured calcareous or dolomitic mudstones with occasional beds of salt or anhydrite. In particular, there are two thick developments of salt beds in the central part of the Cheshire Basin. The Geological Survey Wilkesley borehole (Fig. 13.2) penetrated a combined thickness of 600 m of salt beds, whilst the thickness in Prees-1 was about 430 m. These, together with associated gypsum beds, were probably deposited in extensive salt lakes in the low-lying Keuper plain. Subsidence of the land, together with repeated phases of evaporation, allowed thick salt beds to accumulate. The last event in Triassic times was a widespread invasion of the sea and the deposition of the thin Rhaetic marine mudstones and limestones (21 m thick in Prees-1). The Prees well was spudded on a small outlier of Lower Liassic shales (Fig. 13.2) of which 598 m were penetrated in the well.

Various estimates have been made of the total thickness of post-Permian strata in the deeper parts of the Cheshire Basin. In the Institute of Geological Sciences Regional Handbook, an estimated thickness of 2,620 m of New Red Sandstone is given for the axial portion of the basin. Certainly this was exceeded in the Prees-1 well (Fig. 13.3) which was drilled during 1972–73 to a depth of 12,500 ft (3,810 m) and penetrated 3,000 m of Permo-Triassic redbeds. An earlier deep boring near Formby, penetrated more than 1,676 m of Permian and Triassic redbeds and then entered the Upper Carboniferous (Millstone Grit).

Irish Sea Basin

Knowledge of the stratigraphy of the Irish Sea and Cheshire Basins has greatly improved in recent years with the release of both onshore and offshore deep exploration well results.[5, 8] Shallow sea-bed coring between the Isle of Man and the mainland coast provides some indication of the lithology of the Permo-Triassic sediments infilling the Irish Sea Basin, and the exposures of the basin margin along the Cumberland and Lancashire coast from Furness to St. Bees Head give a further insight into the stratigraphy of the offshore region. Basal Permian scree deposits and coarse pebble beds rest directly against the high ground of the Palaeozoic Lake District Massif. Seawards, these coarse deposits are overlain by younger shales and limestones (St. Bees Shales and Evaporites), broken by occasional intervals of anhydrite.[8] This suggests a westward transition from terrestrial to marine environments. The overlying succession of St. Bees and Kirklinton Sandstones followed by Stanwix Shale resembles that of the Vale of Eden.

The Irish Sea Basin is a half-graben with the deeper eastern side controlled by northwest-southeast and north–south-striking faults. The release of some of the results of drilling within the offshore Irish Sea Basin[5] allows a comparison to be made with the stratigraphy of the Cheshire Basin (Figs. 13.3 and 13.4). Two wells within block 110/8 penetrated shales and sandstones of Namurian or Westphalian age, reddened beneath the unconformity. Subcrop maps of the pre-Permo-Triassic surface[5] show a broad band of Namurian and Westphalian (Coal Measure) rocks occupying an east-west synclinal tract between the Isle of Man and North Wales.

The Collyhurst Sandstone equivalents (260–280 m thick) offshore, away from the source

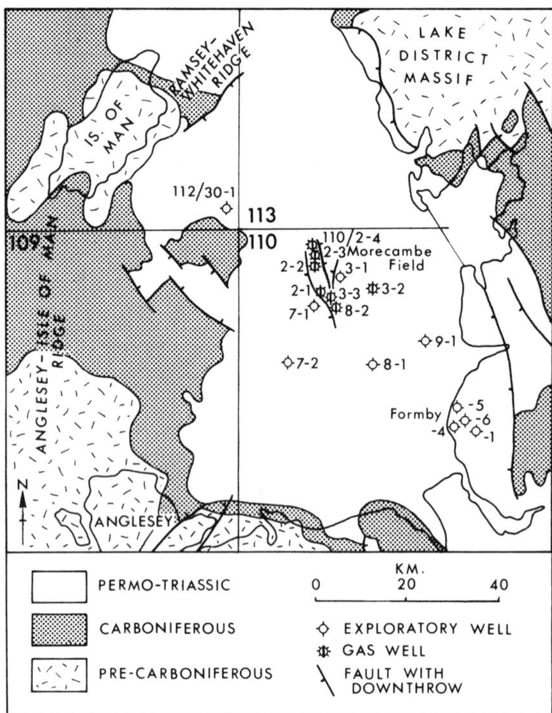

Figure 13.4 Geological sketch map of the Irish Sea Basin; modified after Colter (1978).[8] The junction of Sectors 112–113 to the north and 109–110 to the south is shown.

areas at the basin margins, comprise shales and siltstones. The Gulf/NCB 110/8–1 well (Fig. 13.3) penetrated a shaly equivalent of the Manchester Marl interval, whereas the 110/8–2 well encountered thick salt. Basin margin deposits with supra-tidal and inter-tidal carbonate and anhydrite have been described in boreholes at Whitehaven.[9] The overlying Bunter Pebble Beds of the onshore basin can be traced as far north as the Formby wells but are not represented in the Irish Sea wells. As a consequence there is a thick (1,200 m) unbroken sequence of sandstones with subordinate mudrocks, the Bunter Sandstones (= St. Bees Sandstone) up to the base of the Keuper Sandstone (Hardegsen Unconformity). The Hardegsen Unconformity is thought to represent a considerable change in the depositional environment; the sediments below are interpreted as being derived from the south whilst those above have a predominantly northerly derivation. The fluviatile cross-bedded sandstones of the Keuper Sandstone Formation (200 m) are interpreted as being the product of low sinuosity channel systems, with a middle shalier member produced by higher sinuosity streams and overbank flooding. The unit is distinguished from the underlying sandstones in the offshore wells by use of the gamma-ray log. Porosity is generally good, although, despite the presence of wind-rounded sand grains, dune bedding has not been recognized. There is an upward gradation into the shalier flaggy Keuper Waterstones Formation (70 m) whose top in the Irish Sea is marked by the appearance of salt. A lower variable Saliferous Member (240 m) here passes up to typical Keuper Marl (700 m seen) which crops out on the seabed.

The *Morecambe Gas Field*,[12] in the central part of the Irish Sea Basin (Fig. 13.4), was discovered in 1974 by the Hydrocarbons G.B. Ltd. well 110/2–1. The field is 25 km long, north to south, and 9.6 km wide, with a total area under closure of 114 sq km. Seven further delineation wells were drilled by Hydrocarbons G.B. Ltd. and these penetrated a standard Permo-Triassic redbed sequence. One well encountered reddened pre-Permian (probably Upper Carboniferous) beds. Triassic sandstones provide the gas reservoir, with Keuper Saliferous Beds as a capping series (Fig. 13.3). The reservoir zone is divisible into upper and lower units by porosity-permeability properties. The upper unit comprises the Keuper Waterstones and the upper member of the Keuper Sandstone and contains high porosity sandstones in which porosity has been little affected by diagenetic alteration. A lower unit consisting of the lower members of the Keuper Sandstone and the underlying St. Bees Sandstone (Fig. 13.3) has suffered considerable growth of platy authigenic illite and cementation which considerably decreases permeabilities.[13] Total recoverable gas reserves in the field are calculated at 5 Tcf with about 65% of these reserves within the upper reservoir unit. Production drilling in the field is set to begin in 1982 and the field should be on stream in 1984. One production problem is the shallow depth of the reservoir (3,000 ft: 915 m) and to overcome this innovative slant drilling techniques are being attempted from fixed platforms. The gas will be brought ashore by a 36 inch diameter pipeline to Rampside, near Barrow-in-Furness.

Kish Bank Basin

The basin is a half-grabenal feature elongated northeast-southwest and measuring approximately 48 × 32 km. The sedimentary fill shows a predominant northwestward dip. To the northwest the basin is bounded by the major northeast-striking Dalkey and Lambay Faults, and to the southwest by the Bray Fault which strikes in a northwest-southeast direction (Fig. 13.5). Throws on these major boundary faults are substantial and are often of the order of 3,000–5,000 m.

The basin is cut by the major northwest-southeast Codling Fault, which is probably of post-Jurassic (Eocene or early Oligocene) age. It is a dextral strike-slip fault with a horizontal displacement of approximately 6 km. A number of smaller faults within the basin are parallel to both the Bray and Dalkey-Lambay Faults, while occasional north–south-striking faults also occur. Unlike the major faults, these smaller fractures rarely penetrate the basement and are frequently seen on seismic sections to be absorbed and die out within the salt-bearing section.

Evidence of salt flowage in the probable Keuper succession is seen by the abrupt thickness variations on seismic sections, which also show localized salt diapirism along part of the Codling Fault.

Two wells (Amoco 33/22–1 and Shell 33/21–1) have been drilled to date (early 1982) in the Kish Bank Basin (Fig. 13.5). The results of the Amoco well have recently been published.[14] In addition to this well information sea floor sampling has aided the understanding of the geology of the basin.

The stratigraphic sequence probably comprises Permo-Triassic to Liassic rocks overlying a Carboniferous succession which in turn, as elsewhere in Ireland, rests unconformably on Lower Palaeozoic basement. The latter comprises chloritic slates which may correlate with the Bray Group (Cambrian), or possibly Ribband Group (Cambro-Ordovician) metasediments which crop out onshore and which form the extensive Lower Palaeozoic Leinster Massif of southeast Ireland.

Lower Carboniferous limestones, shales and sandstones crop out extensively onshore north of Dublin. These are likely to extend offshore and to underlie at least the deeper western and northwestern parts of the Kish Bank Basin. They are absent in the southeast of the basin in the Amoco 33/22–1 well. The Lower Carboniferous limestones in turn are probably unconformably overlain by Westphalian to Stephanian sediments.

The Amoco 33/22–1 well[14] encountered an Upper Carboniferous (Westphalian) section 722 m thick, resting directly on Lower Palaeozoic strata (Fig. 13.5). The Westphalian commences with a 10 m basal sandstone horizon which is succeeded by approximately 200 m of Westphalian 'B' cyclothemic siltstones and sandstones accompanied by occasional thin coal seams. This is followed by over 300 m of Westphalian 'C' fine-grained clastics containing a composite thickness of approximately 11 m of generally thin coals with occasional seams thicker than 1 m. The cyclothemic non-marine environment was interrupted during Westphalian 'C' times by a period of sub-aerial exposure and erosion. The Westphalian 'D' succession is approximately 50 m thick and is represented by cyclothemic deposits poor in coal. It is topped by some 150 m or so of possible Stephanian siltstones and shales.

Seismic interpretation suggests that the Carboniferous is unconformably overlain by up to 3,000 m of Permo-Triassic sediments, with the maximum thickness occurring towards the major faults which control the half-graben (Fig. 13.5B). The sequence is probably similar to that seen onshore (Fig. 13.1) in the Kingscourt Outlier northwest of Dublin,[15] and in the Irish Sea Basin.[5] The Permian is likely to be thin and may commence with a Rotliegendes-equivalent continental basal conglomerate similar to the Conglomerate Member of the Kingscourt Gypsum Formation and to the Collyhurst Sandstone of the Irish Sea Basin. This is likely to be overlain by approximately Zechstein-equivalent strata comprising an evaporitic marl sequence similar to the major part of the Kingscourt Gypsum Formation in the Kingscourt Graben and to the Manchester Marl and St. Bees Evaporite Formations of the Irish Sea Basin.

At Kingscourt[15] the Permian is overlain by 400 m of red Triassic sandstones and siltstones (the Kingscourt Sandstone Formation), whereas in the Irish Sea Basin the Triassic sequence comprises 1,400 m of red sandstones overlain by 1,000 m of Keuper marls and saliferous beds. In the Kish Bank Basin the probable Triassic section is of the order of 2,300 m thick. It may comprise some 1,500 m of sandstones (Sherwood Sandstone Group) equivalent to the St. Bees Sandstone, the Lower Keuper Sandstones and the Keuper Waterstones, and is expected to be overlain by an Upper Triassic saliferous sequence equivalent to the Keuper Marl (Mercia Mudstone Group) and the Lower Saliferous Member of the Irish Sea Basin.

Very limited sea-bed sampling and seismic interpretation[16] suggests that up to 300 m of Liassic calcareous sandstones may be preserved immediately adjacent to the down-

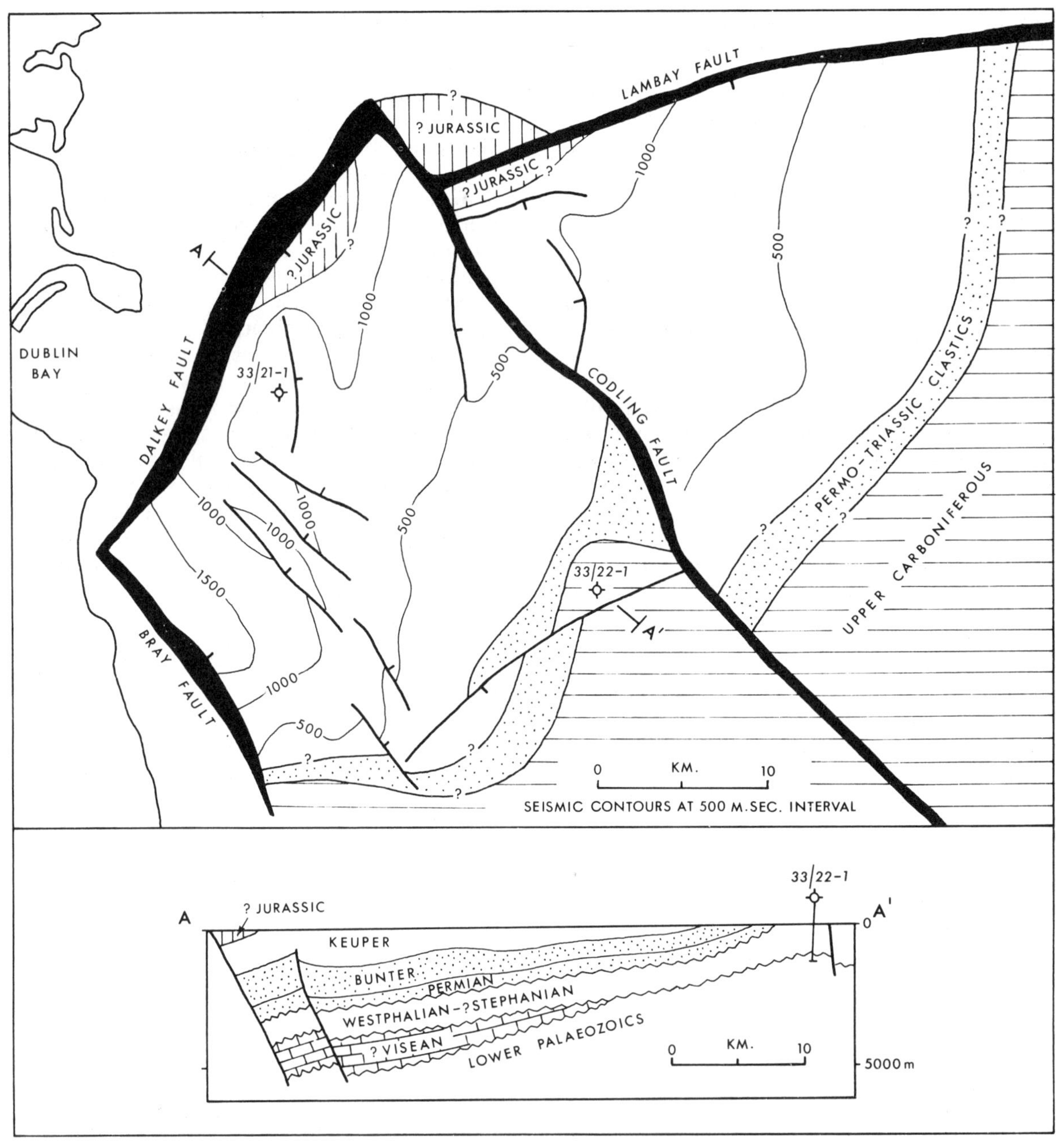

Figure 13.5 Kish Bank Basin.

A. Structure contours (milliseconds) on the base Keuper Salt, with pre-Tertiary solid geology shown outside the area of contouring (after Jenner (1981)[14]).
B. Schematic geological cross-section of the Kish Bank Basin (after Jenner (1981).)[14]

thrown side of a number of the major boundary faults (Fig. 13.5). Lower and Upper Cretaceous sediments appear to be totally absent from the Kish Bank Basin, which is capped by 50–100 m of Tertiary to Quaternary unconsolidated sands.

Hydrocarbon Potential

The search for oil and gas within the *Cheshire Basin* results from geological conditions similar to those in the Southern North Sea. In general, the Permo-Triassic redbed sequence contains few organic-rich source beds. This is also true in the Southern North Sea, where it is generally accepted that the large gas accumulations discovered in recent years in Permian desert sandstones, have their source in the older Carboniferous Coal Measures. A similar situation could exist in the Cheshire Basin. However, in the North Sea, a seal for the Permian gas is provided by the Zechstein (Upper Permian) evaporites immediately overlying the sandstones. No such thickness of Permian evaporites occurs in Cheshire and thus much could hinge on the thickness, nature and extent of the Manchester Marl in the subsurface, and in particular, on its ability to seal any potential trapping situation. A greater potential is present at higher stratigraphic levels where good Keuper and Bunter sandstone reservoirs are overlain by thick developments of salt. The problem here is the long migration route which may have existed between the Carboniferous source for gas and its Triassic reservoir, with increased chances of loss occurring.

Seepages of oil are common in the coalfields surrounding the Cheshire Basin and are found at the surface in several places. For example, Coal Measure sandstones impregnated with tarry oil occur in Coalbrookdale. Of particular interest are the seepages at Formby, near the West Lancashire coast. Here oil was produced for some years from shallow wells drilled in oil-impregnated Keuper Sandstones at the surface. Unfortunately, the deep source of this live flow of oil has so far been sought without success.[10]

Many of the arguments regarding hydrocarbon prospectivity used above for the Cheshire Basin can be applied to the *Irish Sea Basin*. However, despite the development of a halite unit[5] equivalent to the Permian St. Bees Evaporites (300 m thick in the Morecambe Gas Field), the basal unit (Collyhurst-equivalent) comprises red siltstones and mudstones with low reservoir potential. An analogy cannot, therefore, be drawn with the Rotliegendes-Zechstein couple of the Southern North Sea Basin. However, as we have seen, the large (5 Tcf recoverable) Morecambe Gas Field demonstrates the good reservoir qualities of the Triassic sandstones with an overlying seal of Keuper Saliferous Beds. The main potential must therefore rest in locating further structural closures at this stratigraphic level. Some oil prospects fed from Carboniferous sources may exist, but the difficulties of locating and predicting these have been demonstrated onshore.

There are obviously adequate sand reservoirs within the basins in question. However there is variable degree of cementation including silicification and quartz overgrowths.[13] A 500 m thick zone within the Bunter, both in the onshore and offshore wells in the Cheshire and Irish Sea Basins, has low porosity and permeability in the sandstones due to silicification; perhaps due to overburden pressures. Silica cement is also known in the aeolian Penrith Sandstones of the Vale of Eden farther north.

The *Solway Firth Basin* obviously has a lower potential than the larger Irish Sea Basin. However, Coal Measure source rocks are thought to extend northwards beneath the basin from the Ramsey-Whitehaven Ridge and evaporitic potential cap-rocks occur within the Permo-Triassic sequence.

Although the *Kish Bank Basin* is relatively small and two wells have been drilled in the general area, with subsequent relinquishment of the acreage back to the Government, it would appear that the hydrocarbon potential of the basin has not been fully appraised. The first well in the area was located outside the Permo-Triassic basin and was spudded in Upper Carboniferous, while it is generally believed that the second well did not test the complete prospective Mesozoic sequence. The first well did however indicate the presence of potential source rocks. The Westphalian sequence contains medium volatile bituminous coals which were near the peak stage of gas generation. The probable Permo-Triassic sequence is likely to contain abundant clastics, especially in the lower part of the section which by comparison with the Morecambe Gas field of the Irish Sea Basin,[5, 12] might be expected to contain good reservoir rocks with appreciable porosity and permeability. Sandstones are also recorded in the Upper Carboniferous, but tend to be rather thin and to have low porosities and permeabilities, and are therefore unlikely to act as a significant reservoir horizon. The upper saliferous and marly Triassic sequence could be expected to form an adequate seal for any reservoir horizons. This assumes that adequate reservoirs and seals existed when the gas was formed and that gas generation did not take place entirely during the Hercynian episode.

However, the structures in the basin appear from the seismic information to be relatively small. Further problems may also exist in that some of the faults appear to approach, or in some cases actually reach the top, of the

Permo-Triassic section. The exact age of formation of structure and the associated faulting is therefore uncertain. Nevertheless, in view of the proximity of the basin to Dublin, of the shallow water depth and particularly of the apparently favourable geological sequence in the area, it is felt that the basin may hold untested potential for the discovery of economic gas accumulations.

References

1. BOTT, M. P. H. 1964. Gravity measurements in the north-eastern part of the Irish Sea. *Q. J. geol. Soc. London* **120**, 369–96.
2. BOTT, M. P. H. and YOUNG, D. G. G. 1971. Gravity measurements in the north Irish Sea. *Q. J. geol. Soc. London* **126**, 413–32.
3. BACON, M. and McQUILLIN, R. 1972. Refraction seismic surveys in the north Irish Sea. *Q. J. geol. Soc. London* **128**, 613–621.
4. INSTITUTE OF GEOLOGICAL SCIENCES. 1969, 1971. Regional Handbooks, *Central England. Northern England.* H.M.S.O.
5. COLTER, V. S. and BARR, K. W. 1975. Recent developments in the geology of the Irish Sea and Cheshire Basins: *in* Woodland, A. W. (*Ed.*). *Petroleum and the Continental Shelf of North-West Europe, 1, Geology.* Applied Science Publ. London, 61–75.
6. AUDLEY-CHARLES, M. G. 1970. Triassic palaeogeography of the British Isles. *Q. J. geol. Soc. London* **126**, 50–81.
7. WARRINGTON, G., AUDLEY-CHARLES, M. G., ELLIOT, R. E., EVANS, W. B., IVIMEY-COOK, H. C., KENT, P. E., ROBINSON, P. L., SHOTTON, F. W. and TAYLOR, F. M. (1980). A correlation of Triassic rocks in the British Isles. *Special Report of the Geological Society, London, No. 13,* 78 pp.
8. COLTER, V. S. 1978. Exploration for gas in the Irish Sea. *Geol. Mijnbouw* **57**, 503–16.
9. ARTHURTON, R. S. and HEMINGWAY, J. E. 1972. The St. Bees Evaporites – a carbonate-evaporite formation of Upper Permian age in west Cumberland, England. *Proc. Yorkshire geol. Soc.* **38**, 565–92.
10. FALCON, N. L. and KENT, P. E. 1960. Geological results of petroleum exploration in Britain, 1945–57. *Mem. geol. Soc. London* **2**, 56 pp.
11. THOMPSON, D. B. 1970. The stratigraphy of the so-called Keuper Sandstone Formation (Scythian-(?) Anisian) in the Permo-Triassic Cheshire Basin. *Q. J. geol. Soc. London* **126**, 151–181.
12. EBBERN, J. 1981. The geology of the Morecambe Gas Field: *in* Illing, L. V. and Hobson, G. D. (*Eds*). *Petroleum Geology of the Continental Shelf of North-West Europe.* Institute of Petroleum, London, 485–493.
13. COLTER, V. S. and EBBERN, J. 1978. The petrography and reservoir properties of some Triassic sandstones of the Northern Irish Sea Basin. *J. geol. Soc. London* **135**, 57–62.
14. JENNER, J. K. 1981. The structure and stratigraphy of the Kish Bank Basin: *in* Illing, L. V. and Hobson, G. D. (*Eds*). *Petroleum Geology of the Continental Shelf of North-West Europe.* Heyden & Son Ltd, London, 426–431.
15. VISSCHER, H. 1971. The Permian and Triassic of the Kingscourt Outlier, Ireland. *Spec. Pap. geol. Surv. Ireland* **1**, 114 pp.
16. DOBSON, M. R. and WHITTINGTON, R. J. 1979. The geology of the Kish Bank. *Q. J. geol. Soc. London* **136**, 243–249.

Chapter 14

Palaeogeography

From Precambrian to Recent times Northwest Europe has undergone a complex sequence of geological events during which a variety of frequently diachronous structural provinces and dynamic depositional centres were created and deformed. The majority of the geologically young sedimentary basins described in the previous chapters are located in the offshore regions, with the geological sequences in the adjacent onshore areas very often being significantly different. Despite these marked differences between the onshore and offshore regions the familiar shape and coastlines of Britain and Ireland have only emerged in the relatively recent past. This serves to illustrate that the complete geological history of the offshore areas can only be fully understood when looked at in the context of the regional tectonic and palaeogeographic evolution of the entire Western Europe – North Atlantic province. In this chapter we attempt to give a generalized overview, through geological time, of the palaeogeographic evolution of Northwest Europe while concentrating in particular on Britain and Ireland.

The results of many years of exploration throughout Northwest Europe have firmly established that the rocks most likely to contain substantial quantities of hydrocarbons are those which were deposited in Upper Palaeozoic, Mesozoic and Tertiary times. As such, the palaeogeographic evolution of the area from Devonian times onwards is considered in more detail than that of earlier geological times. From Upper Palaeozoic times onwards substantial quantities of sediments were deposited in basins largely created and controlled by the effects of a number of major phases of earth movements. These are as follows:

1. *Caledonian* orogenic (mountain building) events, of Cambrian to Devonian age.
2. *Hercynian* (Armorican) orogenic events, of Upper Devonian to Upper Carboniferous age.
3. *Cimmerian* tectonic events, of Lower Jurassic to Lower Cretaceous age.
4. *Alpine* tectonic events of Tertiary age.

Building of the initial structural framework in Northwest Europe goes back beyond the Caledonian events and began in Early Precambrian times, roughly 2,600 million years ago. The ancient rocks which formed the stable platforms of the Laureatian-Greenland and Fennoscandian-Baltic shields[1] were folded, deformed and metamorphosed. Remnants of these early deformations can still be recognized in the Outer Hebrides and in parts of the northwest coast of Scotland, together with isolated inliers scattered throughout England and Wales.

During the long Precambrian and Early Palaeozoic Eras, a broad sedimentary trough existed between the already rigid shield areas to the east and to the west. This provided a depositional centre into which the erosional products of the flanking shields were poured. The southern extremity of the trough stretched across Scotland, northern England, Wales and much of Ireland. With the gradual collison of the shield areas during Ordovician, Silurian and Devonian times, the powerful fold movements of the *Caledonian* orogeny crumpled these basinal sediments and uplifted them into a chain of high mountains. This fold belt, studded with syn- and late-orogenic extrusive and intrusive igneous rocks, stretched from Norway across the North Sea and as far south as Ireland and southern England. An eastern branch of the belt extended eastwards through the Central North Sea, the Low Countries and northern Germany as far as northern Poland. The western part of the fold belt is preserved in eastern Greenland and in the Appalachians of eastern North America. The worn-down remnants of these ancient mountain chains still form important topographic features at the present day.

In Early Ordovician times the Scottish Highlands were slowly uplifted to form a major land region, while the remainder of the British Isles and Ireland to the south were still submerged beneath a shelf sea. With the continued convergence of both shield areas, similar palaeogeographic conditions persisted into the Silurian, although there was a general tendency towards shallowing of the shelf sea and the appearance of an emerging belt of islands through the Lake District and parts of the Irish Sea and St. George's Channel. By Late Silurian and Early Devonian times the intense Caledonian earth movements caused by the collision of the American and European crustal plates reached their climax, creating one vast mountainous continent and forcing the shallow shelf sea to migrate southwards from all but the southernmost portion of Britain.

With the withdrawal of the sea, continental conditions prevailed over most of the newly created land area. Rapid erosion of this emergent high ground, together with Late Caledonian extensional movements, led to extensive downwarping of localized areas. These included a broad continental depression in the vicinity of southern Norway, the Orkneys and Shetlands, and a narrow downfaulted depression (the Midland Valley) across the centre of Scotland which linked the North Sea sedimentary region with that of the Malin Sea and the northern part of Ireland. Within these newly formed basins thick Devonian fluviatile and lacustrine sandstone sequences were laid down. At this time the general shoreline between the northern continental area and the southward-deepening marine areas stretched from southern Ireland across southern England into northern France, with typically marine limestones, shales and thin sandstones being deposited across much of the Celtic Sea, Cornubian Massif and Western Approaches region. Localized calcalkine and alkaline vulcanism was prevalent at about this time.[1]

The emergence of the Late Devonian-Carboniferous *Hercynian* mountain chain through the Brittany-Normandy peninsula and central France changed the entire palaeogeographic pattern of Britain and Ireland. The mountains created a well-defined southern boundary to the existing Devonian Sea, and provided a rich source of continentally-derived sandstones and mudstones which were shed northwards. Initially the uplift of the Hercynian mountains pushed the edge of the sea northwards, and the Early Carboniferous was marked by the encroachment of marine conditions into continental depressions in northern England and Scotland and the deposition of marine limestones over the more stable regions including much of central Ireland and central and southern England, with marine shales in the more rapidly subsiding regions.[2] These latter areas included southwestern Ireland, the Celtic Sea area, Cornwall and the Southern Upland belt extending from Scotland through the northern part of Ireland. During this period much of Wales formed a large island, known as St. George's Land or the Welsh Massif, which persisted from at least Early Carboniferous times. The extent to which this was linked with the Leinster Massif of southeast Ireland across the area of the present south Irish Sea is not know. During Early Carboniferous time, however, the Welsh Massif extended eastwards to link up with the London-Brabant Massif of southeastern England. During this time also most of Scotland continued to stand proud as a major positive area and was flanked along its southern boundary by a broad plain of lagoons, swamps and deltas in which coal-forming conditions locally flourished.

By Late Carboniferous time the areal extent of deep water sedimentation had contracted and such conditions were restricted to a narrow trough flanking the rising Hercynian mountain chain. With the exception of the land areas formed by the Scottish Highlands, the areally reduced Welsh Massif and the then separate London-Brabant Massif, together with other minor positive areas such as the Southern Uplands, interfingering shallow water and coastal swamp conditions prevailed over most of Britain and Ireland. These swamp conditions led to the deposition of locally very thick coal-bearing sequences throughout central and northern England, the Irish Sea, Midland Valley of Scotland, South Wales, southern England and the Southern North Sea, and over wide areas of Ireland. Subsequent crustal disturbances, largely due to the last orogenic pulses of the Hercynian folding and mountain building event, have meant that such Coal Measures are now only preserved over parts of these areas. Nowhere is this more clearly demonstrated than in Ireland, where the previously widespread Upper Carboniferous coal-bearing sequences are preserved only as small isolated pockets.

Towards the end of the Carboniferous and the beginning of the Permian, the swampy conditions of Britain and Ireland largely gave way to a drier, arid or semi-arid continental environment of low-relief desert plains and newly-created localized fault-controlled depressions within which the products of erosion and weathering collected as a series of aeolian and fluviatile clastic deposits. These sediment-gathering depressions are poorly defined, but probably existed in parts of the Cheshire Basin, Irish Sea Basins (including the Kish Bank Basin and the onshore Kingscourt Outlier), Northern Ireland-Southwest Scotland Basins and Porcupine Basins. Deposition may also have occurred in the Rockall Trough area although the absence of wells in the area means that there is no hard evidence to support this hypothesis (see Chapter 9). It is thought unlikely that such Early Permian sediments are preserved in the Celtic Sea or Western Approaches areas. Again, however, lack of data, together with the difficulties in obtaining precise palaeontological control in Permo-Triassic sequences, casts a certain amount of doubt over the apparent absence of Permian strata in these areas. Further east, in the North Sea region, Early Permian Rotliegendes sedimentation took place in the Northern and the deeper and larger Southern Permian Basins, separated by the Mid North Sea-Ringkøbing-Fyn High.[1] To the south and southwest the Hercynian mountains had been largely worn away by this time and were now replaced by low hills in Cornubia and Brittany.

By Late Permian (Zechstein) time, the sea had once more invaded large areas of the Early Permian continent and had flooded the extensive depressional areas. Two almost enclosed seas, partly separated by a central barrier along the line of the Pennines, were created. Throughout Upper Permian times these seas remained shallow, but varied considerably in extent and salinity. This, together with the fact that at this time the general area lay just north of the equator, led to the deposition of sediments ranging from limestones and marls to evaporites and thick salt sequences. To the east, the larger of the two seas, known as the *Zechstein Sea* covered much of the North Sea region and east coast of England, extending eastwards into Holland, Denmark and northern Germany, and probably northwards via a narrow passage between Greenland and Norway into an open Arctic Ocean. The Zechstein Sea was an extension of the older Northern and Southern Permian Basins with the Mid North Sea-Ringkøbing Fyn High now largely covered by shallow seas. To the west, the less extensive *Bakevellia Sea*[2] occupied much of the northern Irish Sea region and the adjacent Lancashire, Cumberland and northern Irish coasts. It probably extended northeastwards through a narrow passage between the northern Irish coast and the Sea of Hebrides into a basin in the Rockall Trough area. With the lack of data in the Rockall Trough there is still considerable uncertainty concerning the existence of fairly open marine conditions in Permo-Triassic times. However, it does seem possible that a passage-way may have extended southwards from the open Arctic Ocean. It is thought unlikely that an arm of the Bakevellia Sea extended to cover the Celtic Sea-Western Approaches area.

During the period of Late Permian sedimentation, the Bakevellia Sea appears to have periodically extended into the Cheshire Basin and Vale of Eden, and may also have connected with the Zechstein Sea to the east. However, truly marine conditions appear to have been fairly short-lived in the Bakevellia Sea and, as the sea shrank towards the end of the Permian, the shallow marine sequences of limestones, shales and evaporites gave way to marginal marine and eventually continental sediments. By the beginning of the Triassic, arid and semi-arid continental conditions had again returned to much of Britain and Ireland and the former marine depressions now became extensive centres for the deposition of coarse fluviatile and flash-flood pebble beds and aeolian sandstones. With the continuous degradation of the upland mountainous areas, the continental basins of West Britain and the Irish Sea became progressively infilled until the basins eventually linked, south of the Pennines High, with the major North Sea depositional centre.

The onset of the Triassic in the West British and Irish areas was marked by intensified rifting and the development of a complex system of grabens and flexure-bounded troughs,[1] which probably formed in response to a period of increased regional crustal extension reflecting the inherent instability of the Pangean supercontinent. These major depressions, including the Celtic Sea-Cardigan Bay, Western Approaches and Porcupine areas, formed the framework of the later Mesozoic and Tertiary major basins and were the centres of thick sandstone deposition during the early part of the Triassic. The Early Triassic was also marked, especially in Britain and continental Europe, by the re-establishment of shallow marine conditions for a brief period before the return of more typical continental conditions.

During Triassic times, global sea levels appear to have risen gradually,[3] causing an increase in the extent of the main Tethys Basin of Central Europe and the gradual burial of the Mid North Sea-Ringkøbing-Fyn High. This resulted in the formation of one large marine basin that extended from the northernmost part of the North Sea as far as southern Poland. The gradual increase in sea levels gave rise to brief, but periodic, invasions by the sea during Late Triassic times into most of the West British and Irish basins. In consequence there was widespread development of playa lake and sabkha facies evaporites and marls offshore in the Celtic Sea and Irish Sea Basins and onshore in Somerset (Fig. 14.1). Marine invasions of the Scottish fault-bounded troughs were rare or even entirely absent, the sedimentation in these areas being largely restricted to fluviatile and lacustrine deposits. During the period from the Permian to the end of the Triassic climatic conditions became somewhat less arid, reflecting a northwards drift of the Northwest European continent from the equator to approximately 30°N.

The end of the Triassic marked the cessation of the arid continental to shallow marine depositional regime which had persisted fairly continuously over Britain and Ireland since the culmination of the Caledonian orogeny. During latest Triassic to earliest Jurassic time eustatically-rising sea levels, accompanied by a phase of minor earth movement, the *Early Cimmerian Phase*, caused rapid and regional marine transgression and resulted in the inundation of large sections of Northwest Europe. The Tethys Sea of southern Europe spread northwards, across the eroded remnants of the Hercynian mountain range, into northern Europe and Britain, to link up with the Arctic seas further north. The seas also transgressed westwards through the Celtic Sea-Fastnet Basin and Western Approaches area, to link up with a transgressive arm from the Bay of Biscay, and then northwards into the Porcupine Basin, a further arm extended northwestwards through the Irish Sea and possibly into the Rockall-Faeroe area. During the advance of the Liassic Sea, a number of upland areas throughout Britain and Ireland remained emergent as extensive positive features. The Irish landmass, with the exception of the northeastern part, remained topographically emergent as did the Cornish Platform and the Welsh Massif. There is some evidence to suggest that the latter may have periodically extended westwards during Liassic times to link up with the Irish Massif thereby creating a partial barrier between the seas in the Malin Sea-Sea of Hebrides region and those to the south in the Celtic Sea basins. The Grampian and Permian Ridges also remained emergent during Liassic times as did the Shetland and Hebrides Platform. During the Liassic, tranquil shallow warm marine conditions extended throughout most of West Britain and Ireland, over which carbonates were succeeded by thick marl and shale sequences as the seas gradually deepened. Occasional deltaic complexes built out into the seas from some of the upstanding positive areas. These were particularly well developed in the Hebridean area west of Scotland, less well developed adjacent to the margins of the Irish and Welsh Massifs and the Cornish Platform, and were probably only locally or intermittently developed in other areas. There is also clear evidence from seismic, well and seabed sampling information that, although the Liassic sediments tend to be preserved only within the discrete fault-bounded basins, marine deposition extended beyond the basins and covered large tracts of Northwest Europe (Fig. 14.2).

Throughout much of Northwest Europe a major rifting phase referred to as the *Mid-Cimmerian Phase*, which immediately preceded the onset of sea-floor spreading in the central Atlantic area, marks the transition from the Lower to the Middle Jurassic. In some areas, especially in parts of the North Sea, this is manifested in the development of major volcanic centres.[1, 4] In other areas, notably parts of the North Sea and probably the Porcupine Basin, a major unconformity is developed. In yet other areas, such as the general Celtic Sea area, shallow marine conditions similar to those which dominated Early Jurassic palaeo-

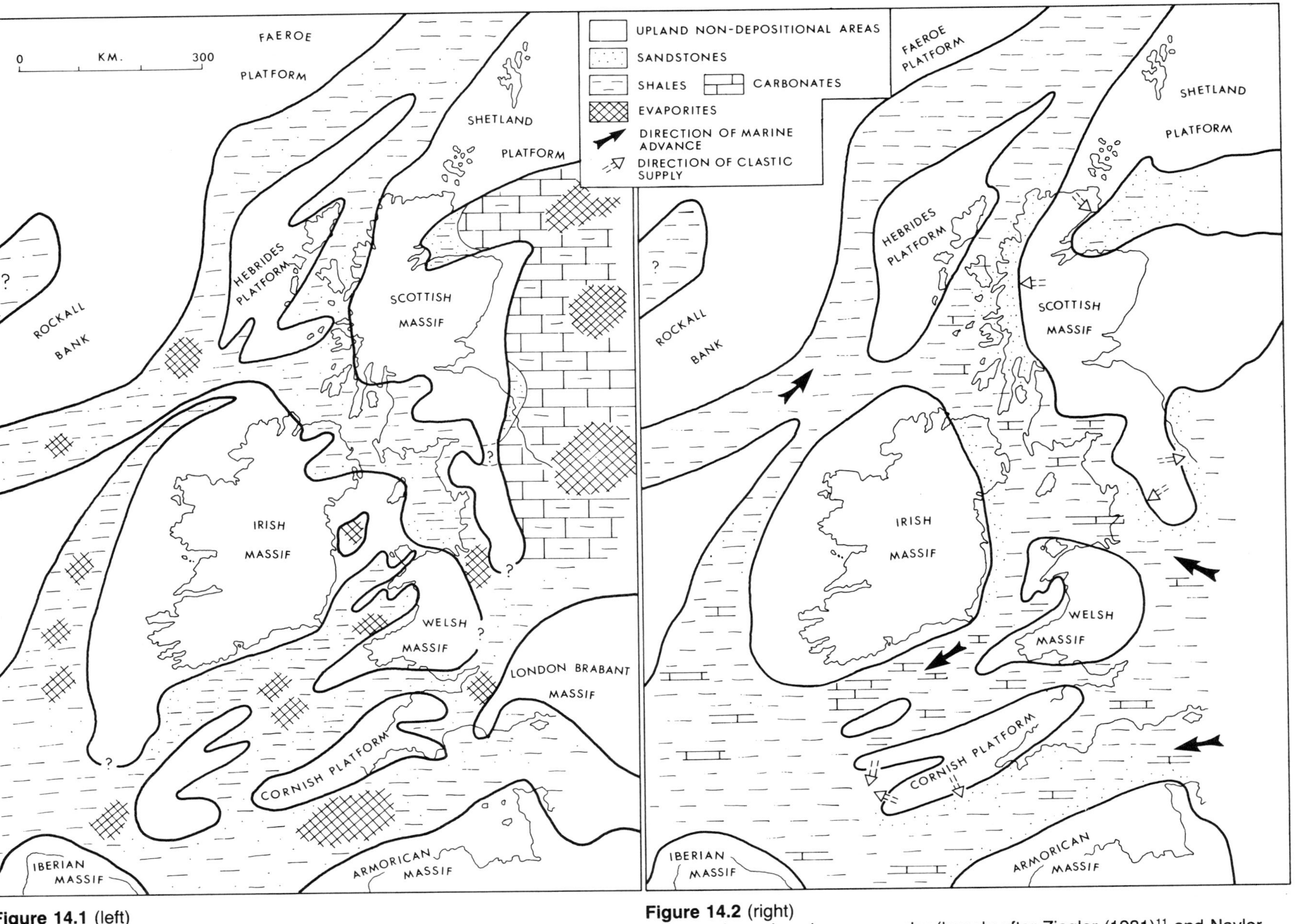

Figure 14.1 (left)
Triassic palaeography (largely after Ziegler (1981)[1] and Naylor and Mounteney (1975)[2]).

Figure 14.2 (right)
Lower Jurassic palaeogeography (largely after Ziegler (1981)[11] and Naylor and Mounteney (1975)[2]).

geography, persisted through Middle Jurassic times with no evidence of an interruption or major change in sedimentation patterns. The only change of significance in that area was that the marine conditions become somewhat shallower during the Middle Jurassic, although there is no evidence of the seas even briefly withdrawing from the area as a whole. However, slight uplift in the general vicinity of the Welsh Massif caused a slight quickening of the shallowing and regression of marine conditions in the St. George's Channel area in end Middle Jurassic times.[2] Marine conditions also appear to have persisted in the general Irish Sea, Rockall and West Scottish areas throughout the Middle Jurassic and it is thought likely that the marine connection between the Rockall and West Scottish Basins with the Celtic Sea Basins, via the Irish Sea area, persisted more or less continuously through this time having been established during the initial Liassic marine transgression (Fig. 14.3).

In the Porcupine Trough marine conditions were completely replaced by continental conditions at the onset of the Middle Jurassic. Extrusive volcanic activity similar to that which occurred in Middle Jurassic times over, for instance, the central part of the North Sea and possibly locally in southern England,[1, 5] is not generally seen in the Irish or West Scottish Basins. Nevertheless some igneous activity at this level is known.[6, 7] This normally took the form of doleritic and gabbroic sills and dykes, which generally appear to be of Late Bajocian age.

The onset of the Late Jurassic was marked by the general regression of the seas from much of the West British and Irish Basins. This probably corresponds to the occurrence of a rifting event in the Atlantic area. In the Porcupine Basin the continental fluvial and alluvial to nearshore conditions which were established in Middle Jurassic times probably continued uninterrupted through most of the Late Jurassic. In the Celtic Sea-Western Approaches area, however, the sudden withdrawal of the seas is reflected in the development of a major unconformity, the *Main Cimmerian Unconformity*, which separates the underlying Middle Jurassic marine sequence from the ensuing continental (alluvial, fluvial and lacustrine) sediments. Further minor tectonic events effected localized incursions and retreats of shallow marine and brackish or transitional conditions in many of the areas, most noticeably in the Celtic Sea area. However, the frequent truncation and extensive erosion of substantial thicknesses of Upper and occasionally Middle Jurassic sediments during Cretaceous and Tertiary times in many of the offshore basins renders the task of reconstructing an accurate palaeogeographic picture for West Britain and Ireland very difficult indeed (Fig. 14.4).

Towards the end of the Jurassic Period the *Late Cimmerian* earth movements began to affect the entire Northwest European area. This major rifting event preceded the onset of Upper Cretaceous sea-floor spreading in the North Atlantic. The major tectonic pulse was accompanied by a significant eustatic drop in sea-level.[3] Throughout Northwest Europe this takes the form of a major regional marine regression.[1, 8] The seas retreated from the entire Irish mainland and from all of Scotland through to Wales and central England. To the east, in the North Sea, shale deposition with only rare clastic influences continued. Marine shale conditions probably persisted in the Rockall Trough while deposition of marine shales and sandstones continued through the Lower Cretaceous in the Porcupine Basin. In the Celtic Sea area the marine to brackish conditions of uppermost Jurassic times gave way to alluvial and fluvial sediments. The onset of the new depositional regime in this area migrated through time, commencing first in the east and central parts of the area, then extending westwards into the Fastnet Basin in early Valanginian time.[6] This probably resulted from preferential uplift in the west of the area due to thermal arching associated with the onset of crustal separation in the Atlantic area to the west.[9] Continental conditions persisted throughout most of the Early Cretaceous and sedimentation during this period was characterized by large river complexes transporting and dumping sediments from the highland areas to the west, north and east, into the restricted depositional centres of the Celtic Sea and southern England. The Early Cretaceous was a rather quiescent time for vulcanism with the only evidence of eruptive activity being in the vicinity of the Cornish Platform and Western Approaches (Fig. 14.5).

During Valanginian to Aptian times sea levels gradually rose and the basins of southern Ireland and England began to receive firstly occasional incursions of brackish and then marine deposits. This transgressive trend was temporarily reversed in parts of the Celtic Sea-Fastnet area in Albian times with a return to marginal continental conditions. These Greensand deposits, clastics and occasional

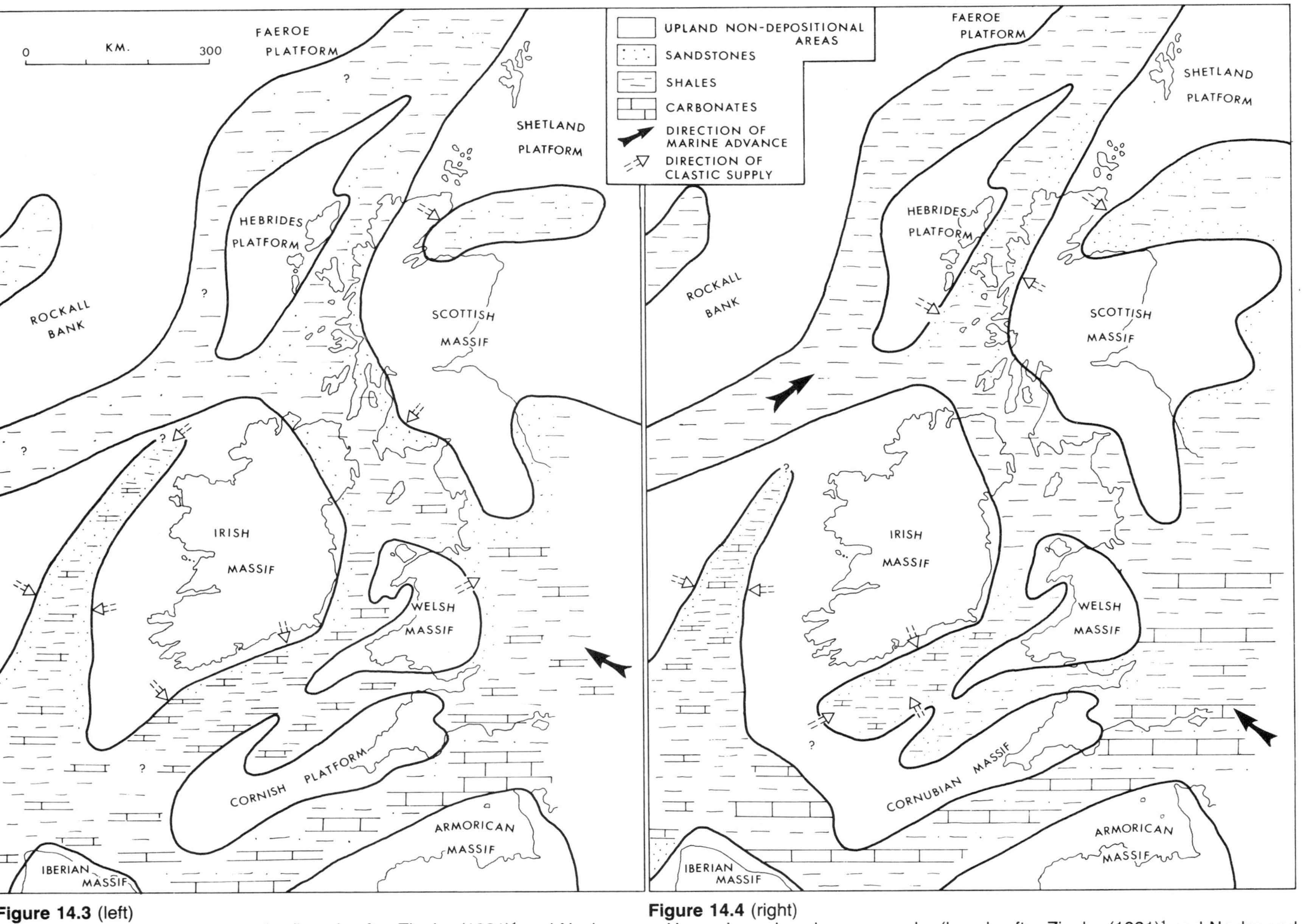

Figure 14.3 (left)
Middle Jurassic palaeogeography (largely after Ziegler (1981)[1] and Naylor and Mounteney (1975)[2]).

Figure 14.4 (right)
Upper Jurassic palaeogeography (largely after Ziegler (1981)[1] and Naylor and Mounteney (1975)[2]).

carbonates, appear to be associated with localized tectonic activity. They are succeeded by a return to marine conditions and, as global sea levels gradually began to rise during the Albian, basin margins were progressively overstepped by a westward advancing sea.

This rise in sea levels was probably induced by the acceleration of the global sea-floor spreading rates[10] and during the Late Cretaceous the sea level occasionally reached levels more than 300 m above the present level.[1, 3] This gave rise to a global transgression. As the sea levels rose throughout Northwest Europe so the clastic influx decreased and this resulted in warm clear water conditions with the resultant deposition of a relatively pure and frequently thick sequence of coccolith-rich chalk. Depositional conditions throughout Northwest Europe during the time probably ranged from inner shelf to bathyal.

These marine conditions, the onset of which was roughly concurrent with the extension of sea-floor spreading into the Labrador Sea and separation of the Greenland-European Plate from the North American Plate, first inundated central and northern England, the Western Approaches, Celtic Sea and Fastnet Basin areas before spreading westwards to link up with the Porcupine Basin and Rockall Trough and extending across the Porcupine and Rockall Banks. How much of the Welsh, Irish, Cornubian and Scottish mountainous areas were submerged during Upper Cretaceous times is still uncertain. However, it is believed that the central cores of these areas remained as positive island features while the chalk sequence was being deposited unconformably across the surrounding shelf and platform regions.

South of the Irish and Welsh massifs a thick accumulation of chalk was deposited from Cenomanian times onwards in the depressions of the Western Approaches, Celtic Sea and Fastnet Basins. Away from these depositional centres the sea spread across the steep, usually fault-bounded, basin margins to deposit a veneer of chalk across the submerged margins of the Cornubian, South Welsh and Irish landmasses. Following inversion and uplift during the Tertiary much of the thin chalk sequence has been removed by erosion and its original extent can only be inferred from a few scattered preserved remnants (Fig. 14.6).

No evidence is seen in the Central Irish Sea area of the deposition of a chalk sequence and it remains conjectural as to whether the Irish and Welsh Massifs which were emergent during the time were connected by the barrier which had periodically existed during earlier periods. Likewise Upper Cretaceous Chalk deposits are absent from most of the Inner Hebrides-West Scottish province, although indications of its former presence in at least the Inner Hebrides region are provided by scattered chalk fragments preserved throughout the area. However, chalk was deposited over the northeastern part of Ireland, in the Rockall Trough and in the basins west of the Shetland and Orkney Islands. In these areas the seas were possibly slightly deeper and the chalk contains more fine-grained clastic material. Throughout Late Cretaceous time vulcanism was only occasionally and locally developed.[1]

Towards the end of the Cretaceous Period the first major (Laramide) pulses of the *Alpine* orogenic episode, centred over southern Europe, began to affect Ireland, Britain and Northwest Europe in general, causing large scale regional uplift and associated local downwarping. The Late Cretaceous Chalk sea slowly began to retreat from the rising landmass of Britan and Ireland withdrawing eastwards and westwards towards the subsiding marginal depressions of the North Sea and Rockall Trough respectively. By Early Tertiary time, the palaeogeography began to assume a shape similar to that of Britain and Ireland today. Land emerged to occupy the entire present-day land surface (with the exception of the extreme southeastern corner of England). Also exposed were large tracts of the surrounding shallow continental shelf in the vincinity of the Shetland and Orkney Islands, to the west of Scotland beneath the Sea of Hebrides and Minches, and further south over the northern Irish Sea, and to the west of Ireland over much of the Porcupine, Rockall and Hatton Banks.

Shortly after the commencement of the Tertiary Period, the continental plate comprising Rockall and Northwest Europe finally separated from the Greenland Plate, and new oceanic crust extended northwards to underlie this part of the North Atlantic. The early phases of plate separation were accompanied by a period of marked crustal extension over much of Britain and Ireland and by the resultant extrusion and intrusion of basic igneous material across northwestern Scotland, the northern part of Ireland, Rockall Bank and Greenland. Some igneous activity also occurred as far south as the western Irish mainland shelf to the east of the Porcupine Basin. This updoming and volcanic activity led to the commencement of erosion and shedding of

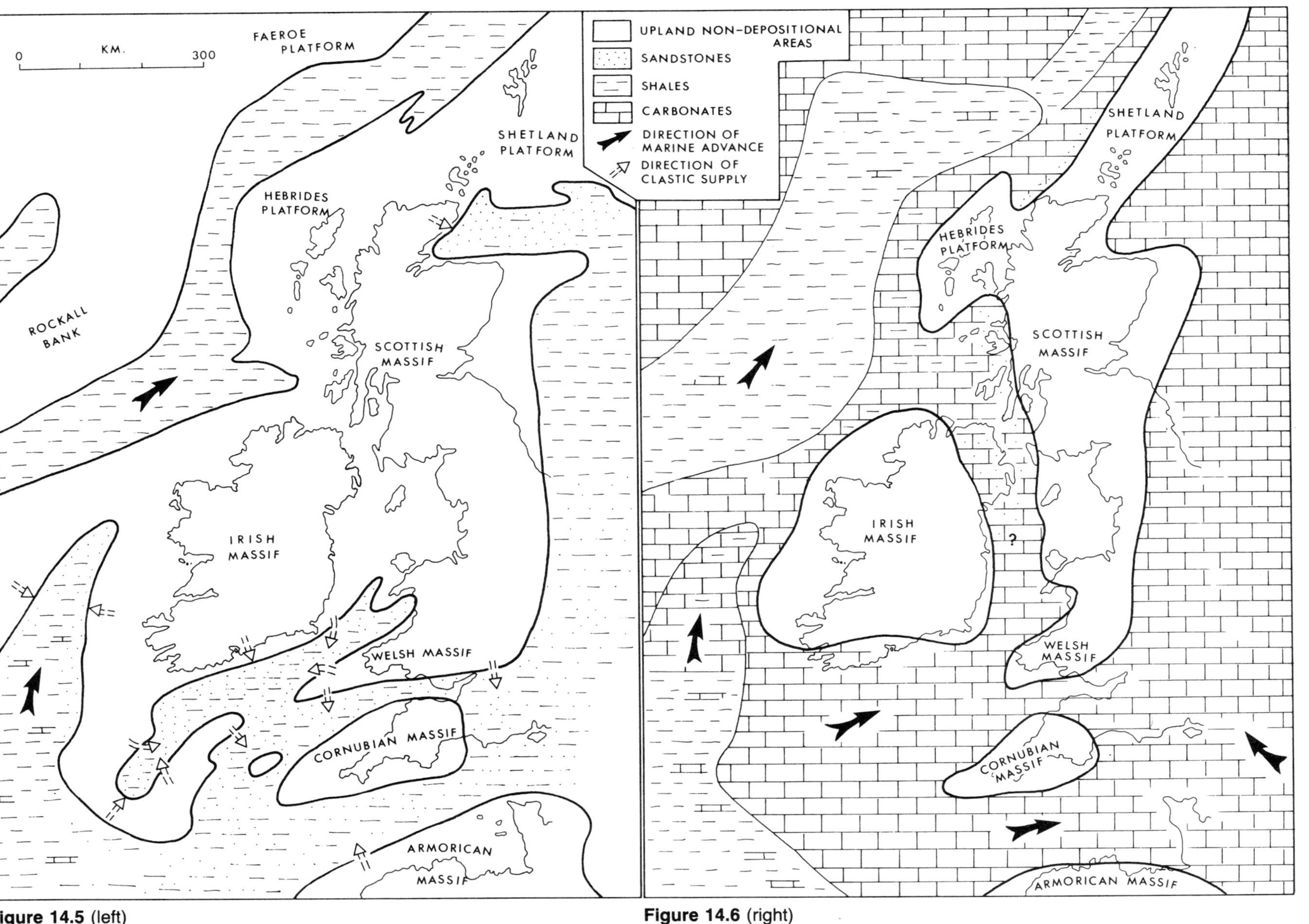

Figure 14.5 (left)
Lower Cretaceous palaeogeography, (largely after Ziegler (1981)[1] and Naylor and Mounteney (1975)[2]).

Figure 14.6 (right)
Upper Cretaceous palaeogeography, (largely after Ziegler (1981)[1] and Naylor and Mounteney (1975)[2]).

Lower Tertiary sediments and volcaniclastics into the adjacent basins. Regional subsidence of the major sedimentary basins, notably the Porcupine, Rockall, Hatton-Rockall and Faroes Basins quickened appreciably in Eocene times as sea-floor spreading became established. The earlier extensional stresses ceased to dominate Britain and Ireland and the structural and sedimentological pattern became gradually influenced by the marginal effects of the developing Alpine mountains to the south. Great thicknesses of terrigenous sediments accumulated in the major sedimentary basins as deposition more or less kept pace with subsidence rates.

In the North Sea area similar thicknesses of Tertiary clastic sediments were deposited. To the south, in southern Europe and the Mediterranean region, the Alpine orogenic movements reached a climax towards the end of the Early Tertiary, throwing the thick marine sediments of the Tethys Sea into huge folds and overthrusts.

Throughout the West British and Irish areas these Alpine movements played very little part in altering the palaeogeographic pattern created by the early stages of plate separation, and the large areas of continental shelf which were emergent throughout most of the Tertiary only became submerged again in the relatively recent Quaternary Period, giving the final touches to moulding the indented coastlines of Britain and Ireland as they exist today.

References

1. ZIEGLER, P. A. 1981. Evolution of Sedimentary Basins in North-West Europe: *in* Illing, L. V. and Hobson, G. S. (*Eds*). *The Geology of the Continental Shelf of North-West Europe.* Heyden & Son Ltd. London, 3–39.
2. NAYLOR, D. and MOUNTENEY, S. N. 1975. *Geology of the North-West European Continental Shelf.* Volume 1. Graham & Trotman Ltd. London, 162 pp.
3. VAIL, P. R., MITCHUM, Jr., R. M., TODD, R. G., WIDMIER, J. M., THOMPSON, S., SANGREE, J. B., BUBB, J. N., HATFIELD, W. G. 1977. Seismic stratigraphy and global changes of sea level: *in* Payton, C. E. (*Ed.*). Seismic Stratigraphy, Application to Hydrocarbon Exploration *Mem. Am. Assoc. Petrol. Geol.* **26**, 42–212.
4. WOODHALL, D. and KNOX, R. W. O. B. 1979. Mesozoic volcanism in the northern North Sea and adjacent areas. *Bull. geol. Surv. G. B.* **70**, 35–55.
5. HALLAM, A. and SELLWOOD, B. W. 1968. Origin of Fullers Earth in the Mesozoic of Southern England. *Nature* **200**, 1193–1195.
6. ROBINSON, K. W., SHANNON, P. M. and YOUNG, D. G. G. 1981. The Fastnet Basin: An Integrated Analysis: *in* Illing, L. V. and Hobson, G. D. (*Eds*). *The Geology of the Continental Shelf of North-West Europe.* Heyden & Son Ltd., London, 444–454.
7. CASTON, V. N. D., DEARNLEY, R. K., RUNDLE, C. C. and STYLES, M. T. 1981. Olivine-dolerite intrusions in the Fastnet Basin. *J. geol. Soc. London* **138**, 31–46.
8. ZIEGLER, P. A. 1978. Northwestern Europe: tectonics and basin development. *Geol. Mijnbouw* **57**, 589–626.
9. ALLEN, P. 1981. Pursuit of Wealden models. *J. geol. Soc. London* **138**, 375–405.
10. PITMAN, W. C. 1978. Relationship between eustasy and stratigraphic sequences on passive margins. *Bull. Geol. Soc. Am.* **89**, 1389–1403.

Chapter 15

History of Oil and Gas Exploration

Introduction

The existence of many of the Mesozoic-Tertiary basins described in this book was unsuspected only two decades ago. Since then there has been a tremendous increase in the geological knowledge and understanding of these areas, and the enormous financial investments by the exploration companies have been rewarded in a number of the areas by the discovery of significant accumulations of oil and gas. Several reasons exist for the explosion in offshore interest and knowledge and these form an interwoven thread of progress which is the subject of much of this chapter.

Throughout this century there has been an increased impetus to explore and carry out research in the seabed areas of the world. The pace of exploration has quickened to almost a frantic gallop in the last decade or so in response to increased demands by the industrialized countries, and the realization or fear of the inherent instability in the continuity of supply from many of the traditional oil exporting regions of the world. This quickening of exploration and research has been aided considerably during the past thirty years by dramatic improvements in equipment for carrying out geophysical surveys and particularly high quality deep seismic reflection work. The oil industry, continually driven by the need to establish new reserves, has focused its attentions and efforts on exploration and production in areas of increasingly greater water depths. This, in turn, has been made possible by the construction of massive, expensive and highly technologically complex drilling rigs and platforms.

The major thrust in the offshore search for hydrocarbons in Northwest Europe was prompted by the discovery in 1959, near Groningen in northern Holland, of vast reserves of natural gas in Permian aeolian sandstones at a depth of 3,000 m. The source of the gas is almost certainly the Upper Carboniferous Coal Measures, now known to underlie some of the other basins described in this book.

The Groningen discovery focused attention on the possibility of oil and gas accumulations existing beneath the North Sea. However, at this time no agreement had been reached regarding the subdivision of the North Sea between the surrounding countries. The Geneva Convention of 1958, which was to act as the basis for this agreement, had not yet been ratified by sufficient countries to bring it into force. Nevertheless by 1963 several of the bordering countries were enacting legislation covering the award of concessions, operating procedures and taxation.

In the meantime exploration became more active. In 1962 Shell, B.P. and Esso carried out a joint reconnaissance seismic survey of the east coast of the U.K. between Lowestoft and the Firth of Forth. A similar regional programme was carried out in German waters by a consortium headed by Mobil. Other regional programmes were instigated in 1963, and in addition much of the North Sea was covered by an aeromagnetic survey during the same year.

In May 1964 the U.K. ratified the Geneva Convention of 1958, this being the 22nd and last ratification required to put the Convention into effect. Later in the same year agreement was reached by which the North Sea was subdivided into seven sectors, each sector being allocated to the respective bordering nation. The United Kingdom passed the Continental Shelf Act (effective 15 April 1964), designated a portion of their North Sea sector and later in the year issued 52 licences. These covered parts of the Southern North Sea, although there was particular interest in blocks covering the Mid-North Sea High. The first North Sea well, Nordsee B-1, was spudded in German waters in May 1964, and the first U.K. well was Amoseas 38/29–1, spudded in December 1964. The first commercial discovery was made in 1965 by B.P.'s 48/6–1 well (West Sole Gasfield). Further drilling demonstrated that the Southern Basin was highly prospective whilst the poor results on the Mid-North Sea High led gradually to relinquishment of acreage on that structure. The Southern North Sea successes pushed exploration first northwards into the Central and Northern North Sea where the first oil discovery, the Ekofisk Field in December 1969, was followed by many more, and then westwards to the deeper water basins off western Britain and Ireland. Attention was also focused on the southern basins between England and France.

The North Sea agreements did not resolve the problem of the offshore boundaries in the south and west between the Republic of Ireland, Britain and France and these remained in dispute. As a result exploration was delayed in the offshore zones affected by the dispute. However, the division of the Channel-Western Approaches between the United Kingdom and France was settled by arbitration in 1976, thereby paving the way for designation and licensing by the respective governments, and then exploration by a variety of companies. The division of the shelf still remains to be resolved between the British and Irish governments, but the parties have agreed to go to arbitration.

We have seen in the previous chapters that many of the first probings of the offshore areas west of Britain and Ireland were made by university and government research teams. This work has continued and some aspects of exploration, such as bottom sampling, are not normally repeated by the oil industry. With awakening interest in the areas west of Britain, oil companies acquired high quality seismic data, normally too expensive to be shot by university teams. The first speculative seismic survey, making data generally available to the industry, was shot in 1970 in the area west of Shetlands. The speed with which the industry has moved since then may be judged by the fact that hundreds of thousands of kilometres of deep seismic data have now been acquired in the areas west of the British Isles and Ireland. The result is that even in the unlicensed areas there is a great deal of detailed knowledge now available to the oil industry. Thus, at an early stage in exploration history it was possible for a company, and also a government, to take an overview of a whole region. This is in contrast to exploration in the past onshore in continental areas where exploration has tended to work gradually into a sedimentary basin, and where it has taken several decades for the overall basin architecture to become apparent. This, in turn, has had an impact on the philosophy of exploration.

The procedure for allowing and containing offshore exploration is similar throughout most of the countries of Western Europe. Initially an area of water is designated by the government of the country concerned, which effectively takes that water under the jurisdiction of the government for the purpose of exploration. It is normally possible at this stage for companies to acquire non-exclusive licences which allow for the exploration of, but not production of, petroleum. It is during this phase that wide-grid regional seismic data are usually obtained. This may be done by exclusive one-company shooting, by groups of companies acting together or by purchase of speculative seismic data from seismic contracting companies. Normally, the following stage is for the government to offer and grant licences on offshore blocks within the designated area. A company then has exclusive rights on the block for a period of several years. During this phase, the company would then acquire more detailed seismic coverage over any interesting subsurface structures and eventually drill a test well. In the event of a discovery, then it is possible for the company concerned to convert the exploration licence to some form of production licence which allows for production of petroleum from that field over a period of years.

Since the first offshore exploration for oil on the Northwest European Shelf in the early 1960s the pattern of exploration and production has differed from country to country. This is true also of the three countries which border the western seas under discussion; namely the United Kingdom, the Republic of Ireland and France. The progress in the search for offshore hydrocarbons will be described for each in turn. For historical purposes an account of relevant onshore work is also given.

United Kingdom

There is a long history of interest in, and exploration for, hydrocarbons onshore in Britain. Natural seepages of oil at the ground surface have been recognized for centuries in different parts of Britain in rocks of various ages. Seepages have been recorded in places as far apart as Coalbrookdale in Shropshire, and Lulworth Cove in Dorset, although by far the most common local occurrence of oil was associated with the collieries in the north of England. The volumes of oil involved in surface seeps are usually very small.

Natural gas was accidentally discovered whilst drilling a water well at Heathfield in Sussex in 1895, and until recently was used to light the local railway station. The demand for petroleum during the period of the First World War prompted the British Government to seek indigenous oil supplies and a number of wells were drilled after the war with limited small success. From that time up until the last ten years, most of the exploratory drilling in the United Kingdom was carried out by British Petroleum. In the period 1936 to 1968 about 20 small fields were discovered with the relatively insignificant cumulative production of 15 million barrels of crude oil, by comparison with the *daily* consumption of 2 million barrels of oil in Britain at the present time. Many of these small onshore fields occur in the Carboniferous, and particularly the Coal Measure, rocks in the north of England. However, interesting fields also produce from Jurassic rocks in the Kimmeridge – Wareham area of Dorset. Although a number of the older fields are now depleted and no longer produce oil, the phenomenal success of the offshore North Sea area, combined with the rising prices and scarcity of oil, has led to a new phase of licence awards and resultant drilling activity onshore in Britain in which deeper rock horizons are likely to be tested. The interesting discovery of significant quantities of oil at Wytch Farm near the south coast of England and promising results in a number of other areas has added impetus to this search.

As described above, the first major exploration thrust in U.K. offshore areas was in the North Sea and this has continued to the present day. Nevertheless, the first offshore exploration well in British waters was drilled in 1963 by B.P. (Lulworth Banks No. 1) in the English Channel to a depth of 762 m. The well was approximately 13 km west of the small onshore oilfield at Kimmeridge which produces from Jurassic limestones and fractured shales.

Ever since the commencement of offshore exploration operations in the United Kingdom, the geographic location of the exploration effort has been decided and guided by the Government by means of a succession of licensing rounds. Following the introduction of the Petroleum (Production) Continental Shelf and Territorial Sea Regulations, 1964, it was announced that applications for production and exploration licenses would be received. Exploration licences grant the holder the non-exclusive right, on areas not covered by production licences, to carry out exploration surveys in addition to shallow stratigraphic drilling. Production licences grant exclusive rights to search for, and produce petroleum from, certain blocks. Production licences are valid for an initial six year period, after which, following relinquishment of 50% of the licensed area, they may be renewed for a further forty years.

A total of 960 blocks, all in the North Sea, were offered for licence and production licences were subsequently granted on 348 of these. Each of the 53 licences carried a commitment to drill at least one well. This series of awards is known as the U.K. First Licensing Round.

The Second Licensing Round took place the following year, 1965, when some 1,102 blocks were offered. The most significant difference between this and the preceding Round was that, in addition to blocks in the North Sea, a number of blocks were offered in the English Channel and the Irish Sea Basins. A total of 127 blocks, comprising 37 licences, were awarded and although no blocks were taken up in the English Channel, 5 blocks were licensed in the Irish Sea Basin. The first well on one of these blocks, 110/8, was drilled by the Gulf Oil/National Coal Board consortium in 1969 and this marked the first exploration drilling operation west of the British mainland.

The Third Licensing Round was announced at the end of 1969 when a total of 157 blocks were offered in the North Sea. The precedent established in the preceeding Round was again followed with a considerable number of the blocks on offer being in the waters west of the

British mainland. Another one of the significant features of this Third Round was that the Government made it clear that for these blocks west of Britain some form of either partnership or option with the National Coal Board or the Gas Council was a necessary precondition for applications. Some 106 blocks were subsequently licensed in 1970, and of these 13 lay in the Irish Sea Basin, 5 in the St. George's Channel Basin and 4 were to the west of the Orkneys.

In 1971 additional areas of the U.K. Shelf were designated by the Government. These covered the northern part of the North Sea together with large areas west and southwest of Britain.

Also in 1971 a further 421 blocks were offered for licence under the Fourth Licensing Round. However, unlike any of the previous Rounds, a further 15 blocks were offered on a competitive cash tender basis. The blocks on offer in this Round included previously relinquished acreage, previously designated areas which had not been licensed before and parts of the newly designated areas. Again the trend established in the previous rounds continued with increasingly more blocks on offer in areas other than the North Sea. A total of 282 blocks were subsequently licensed. Of these, approximately 50 blocks lay in the Orkney and West Shetlands Basins, approximately 40 were in the South Celtic Sea/Bristol Channel/St. George's Channel Basins while a further 3 blocks were licensed in the Irish Sea Basin. Fourth Round licences led to the drilling in 1972 of the first well, with Esso as operator, in the West Shetlands Basin. A further 13 wells followed on Fourth Round blocks in this area before the B.P. 206/8-1A discovery well in 1977, which paved the way for the later appraisal of the large Claire Field containing low gravity oil. Drilling on Fourth Round acreage in the Irish Sea Basin also led to the discovery of the Morecambe Gas Field. The discovery well, 110/2-1, was drilled by Hydrocarbons G.B. Ltd in 1974.

Also in 1974 the U.K. Government made a further designation of acreage to the west of Scotland. The result of this was to extend the U.K. designated territory westwards across the Anton Dohrn and Rockall Banks.

The Fifth Licensing Round of 1976/1977 offered a total of 71 blocks, again in a variety of areas. As with the previous round, a number of the blocks on offer were blocks which had been relinquished in earlier rounds. A feature of this Round was that the blocks were offered on condition that the British National Oil Corporation or one of its subsidiaries be a co-licensee entitled to a 51% share of all licences. Of the blocks which were west of the British mainland 7 were in the English Channel Basin, 5 in St. George's Channel Basin, 1 was in the Irish Sea Basin and 4 lay to the west of Scotland.

Following Arbitration the median line between Britain and France was agreed in 1976. Two periods of designation followed this in the British sector, one in 1977 and the final one in 1978, particularly affecting the Western Approaches area.

A number of these newly designated blocks were included in the Sixth Licensing Round of 1978/1979 where a total of 46 blocks were offered and 42 were licenced. Of those licenced in the area west of Britain, 14 were in the Western Approaches Basin and about a dozen lay to the north and west of the Shetlands. As with the preceding Round, the British National Oil Corporation had the option to take 51% of any licence. The first of those Sixth Round blocks in the Western Approaches, 73/7, was drilled during 1981 by Phillips.

The Seventh Licensing Round of 1980/1981 offered an initial 80 blocks for licence. In response to requests from the industry a further 42 blocks were later offered in the Northern North Sea, with these blocks to be nominated by applicants and with a premium of £5m per block to be paid by successful applicants. Under the terms published in connection with the Round the British National Oil Corporation had the option to purchase up to 51% of any licence. The discretionary blocks were offered in the Moray Firth, in the Southern North Sea, in the hostile waters north of latitude 62°N in the northern part of the North Sea, north of the Shetland Islands and in the Western Approaches and English Channel Basins. In addition to the 42 chosen premium blocks, 53 blocks were licensed of which 16 were in the English Channel/Western Approaches Basin.

Ireland

The evolution of the pattern of exploration for hydrocarbons in the Republic of Ireland is quite different from that of Britain. Ireland lacks the extensive history of exploration for onshore oil and gas which is a feature of its geographic neighbour. Oil seepages and even the smallest of oil fields are unknown onshore in Ireland, but perhaps the most significant single difference between the countries is that

Mesozoic rocks, which are the source and reservoir for much of the onshore hydrocarbons in the U.K., are almost totally lacking on the mainland of the Irish Republic.

Exploration for hydrocarbons in Ireland began in 1959 when the Ambassador Irish Oil Co. received exploration rights for 20 years for the entire country, together with all offshore areas under Irish jurisidiction, subject to surrender of one quarter of the original area each 5 years. Conoco and Marathon Petroleum Ireland Ltd farmed into the licence in 1961, and in 1962 to 1963 some 6 exploration wells were drilled by the consortium to test onshore Carboniferous prospects. While four of these were dry, the other two encountered small quantities of gas. However, following this initial onshore programme interest faded in onshore prospects as attention began to turn to the offshore regions following the early successes in the Southern North Sea. Within the past few years a slight revival of interest has occurred with the main area of attention focusing on the Northwest Carboniferous Basin where the earlier indications of gas were encountered.

In 1964 Conoco withdrew from the licence, with Ambassador following in 1966, leaving Marathon as the sole licensee. In 1968 the Irish Government made its first offshore designation of territory (Fig. 15.1) claiming an area slightly larger than the acreage held by Marathon. The following year a revised agreement was negotiated between the Government and Marathon under which the company relinquished any claims it might have had to the Continental Shelf with the exception of the areas it already held. Also in 1969 the first offshore exploration seismic survey took place in the Celtic Sea.

In 1970 the Irish designated area was further extended to cover more of the inner continental shelf and extend the area as far west as to include the northern part of the Porcupine Seabight. Then, on 16th May 1970, the first well (48/25-1) in the Irish offshore was spudded in the North Celtic Sea Graben with Marathon as operator. This well was plugged and abandoned early the following year. The company then drilled another unsuccessful well and immediately following this returned to drill a second well on block 48/25. This well, completed in late 1971, was the discovery well of the Kinsale Head Gas Field.

Also in 1971 Marathon concluded a farm-out agreement with Esso Exploration Inc. under which the latter acquired a 50% interest in some 50 blocks in the western part of the Celtic Sea. Exploration continued in the Celtic Sea throughout the 1970s with a significant number of wells being drilled by both Esso and Marathon. However, although no further commercial discoveries were made during that time, the companies were granted a total of 20 Petroleum Leases covering 37 blocks which allow production to be carried out for a period of 21 years. (Production Leases awarded on blocks licensed under the 1975 Licensing Terms and Conditions will be for 28 years.)

Offshore exploration took a further significant step forward in 1971 with the issue of a number of Petroleum Prospecting Licences which granted non-exclusive rights to carry out exploration surveys in any areas not covered by Petroleum Leases or Exploration Licences. As a result a number of other companies began to carry out reconnaisance seismic shooting in areas such as the Fastnet Basin, the Porcupine Basin and the Kish Bank Basin.

In 1972 two further designations of offshore acreage took place. The first included the remaining part of the Porcupine Seabight and carried the designated area into the deeper waters west of the Porcupine Bank. The second designation extended the area across much of the Rockall Trough (Fig. 15.1).

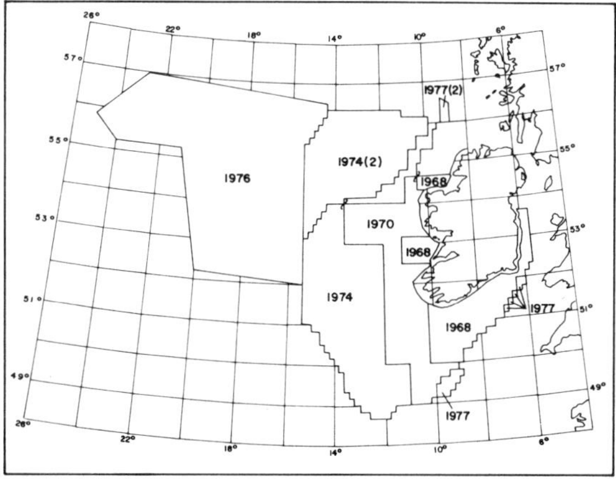

Figure 15.1 Progressive stages of offshore designation, Ireland.

In 1975 the Department of Industry and Commerce published a set of new Terms and Conditions under which exclusive Exploration Licences would be awarded. The current legislation governing exploration for hydrocarbons is contained in the Petroleum and Other Minerals Development Act, 1960, and the Continental Shelf Act, 1968. Applicants were invited to apply for all blocks within Irish designated territory with the exception of blocks held by Marathon and Esso/Marathon and 24 other blocks which at the time were being negotiated for by a number of groups of companies. In 1976 the first Exploration Licences were granted to 11 different groups for a total of 43 blocks. The licences granted the holders exclusive rights to carry out exploration, including drilling, on the specified blocks. Licences were issued for an initial 6 year period where water depths are less than 600 ft, or 9 years where they exceed this figure. A 50% relinquishment is required after 4 or 6 years, depending on the length of the licence. In this, the First Irish Licensing Round, successful applicants were awarded blocks in a range of offshore basins, including the Fastnet, Porcupine, Northwest Irish Offshore and Kish Bank Basins. Although this award of licensing marked the beginning of a more flexible licensing policy it also heralded more demanding and realistic terms and conditions. Significantly, the new terms made the provision for state participation (on a fully paying basis) in the event of a commercial discovery.

In 1976, immediately following the First Round awards, the first drilling outside the Celtic Sea Grabens took place when three wells were drilled in approximately 130 m of water in the adjacent Fastnet Basin. Also in 1976 a further major designation took place extending the area westwards across the southern part of the Faeroe Plateau. The most recent designations took place in 1977 when firstly two small areas were designated in the Celtic Sea and secondly in a small area between the northwestern part of Ireland and the Anton Dohrn Bank. That same year saw further drilling taking place in the Fastnet Basin, while the first well was drilled in the Kish Bank Basin and also in the deep waters of the Porcupine Basin when Shell drilled 35/13–1 in 482 m of water.

The next significant development in the Irish offshore licensing policy came in 1978 when the first of a number of 'option' agreements were signed with exploration companies. Under the terms of an option, the holder is required to carry out a specific work programme, normally comprising seismic shooting, and, following a certain time span (usually 12–18 months), is obliged to decide if a certain number of the blocks are to be kept in which case an Exploration Licence is committed to. Almost all option agreements have had signature bonuses paid to the Government, usually at a rate of approximately Ir £60,000–100,000 per block.

Since the First Round in 1976 the Irish licensing authorities have adopted an 'open door' policy with respect of licensing. Companies are free to approach the Department of Industry and Energy with proposals to licence any specified blocks in open acreage. This policy has proved to be quite successful in that it ensured the granting of a fairly regular number of new Exploration Licences while allowing the oil companies to chose the areas which they felt held the best potential.

In November 1980 the Irish Department of Energy announced its intention of holding a Second Licensing Round, with the closing date for receipt of applications being 29th January, 1981. In all, 108 blocks were offered, and of these 45 were in the North Celtic Sea Graben, 46 in the Porcupine Basin, 7 in the Slyne Trough, 6 in the Donegal Basin and 4 in the Kish Bank Basin. However, in keeping with its policy of flexibility concerning licensing arrangements, and bearing in mind that most of the offshore Irish basins still represent relatively frontier and expensive exploration acreage, the Irish authorities also announced that its 'open door' policy would still remain in force for acreage not included in the Second Round. A total of 35 companies were involved in applications under the Second Round on behalf of 10 consortia. At the time of writing (early 1982), negotiations were in progress between the Dept. of Industry and Energy and the various applicants and it was generally expected that awards would be announced in June 1982.

France

The exploration and licensing history in the French portions of the Channel and Western Approaches has been significantly different from that in United Kingdom and Irish waters. Exploration in France began in the 1950s and led to the discovery by Esso of two oilfields, Parentis and Cazaux, in the onshore Aquitaine Basin. The first offshore seismic work began in the early 1960s and this highlighted three main areas of potential interest; the Gulf of Gascogne in the Bay of Biscay southwest of Bordeaux, the Gulf of Lion in the Mediterranean Sea and the Mer d'Iroise off the coast of Brittany.

As far back as 1961, and following extensive seismic shooting in the area, Esso, as operator for a three company consortium, had applied for offshore acreage in the Gulf of Gascogne, effectively an extension of its coastal territory. The consortium was granted its first offshore permit in 1964. The first of a number of unsuccessful wells was drilled in 1966.

Then in 1968 drilling commenced in the second area of interest, the Gulf of Lion, with CFP as operator. As with the Gulf of Gascogne the results were disappointing.

Since 1964 a number of companies had expressed interest in the newest of the three offshore exploration areas, the Mer d'Iroise. However, in 1970 the French and British Governments began a debate on the ownership of the seas of the Western Approaches. The dispute went to Arbitration in 1974, and a year later, in April 1975, the French Government awarded the first exploration permits in the Mer d'Iroise. These were awarded to a consortium headed by SNPA and included Total, B.P. and Shell. Three large blocks of territory lying to the west of Brest were involved: Mer Celtique (15,000 sq km), Armor (18,000 sq km) and Iroise (9,000 sq km) (Fig. 15.2). The blocks were granted for a period of up to 15 years, comprising an initial tenure of 5 years, followed by a 50% relinquishment. A 12½% relinquishment is required after a further 3 years. There is also a provision for a 50 year exploitation period. Drilling commenced in 1975 with the Lizzen-1 well, followed immediately by the Lennket-1, both in the Armor prospect. Both wells were plugged and abandoned as dry holes.

Although a small number of other exploration and reconnaissance permits have been granted in the French Atlantic waters in the general vicinity of the Mer d'Iroise, the exploration efforts to date in the area have not been very encouraging. A total of 11 exploration

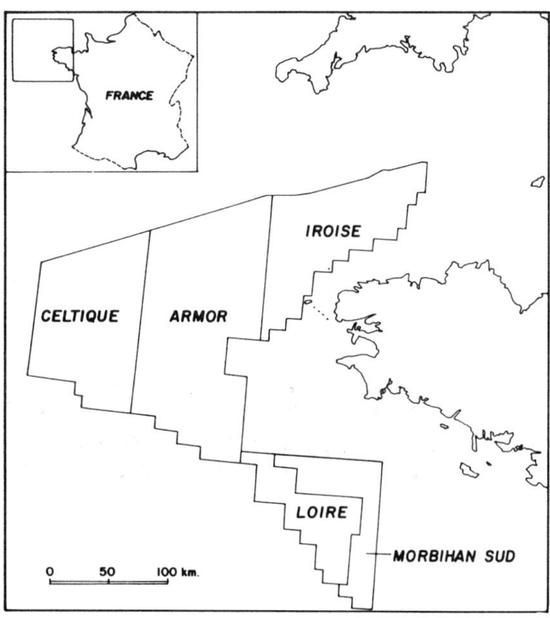

Figure 15.2 French offshore licence areas in 1980, Western Approaches region.

wells have been drilled but all were plugged and abandoned as dry holes. Three wells were drilled in the Mer Celtique permit, 4 in the Armor permit, 3 in the Iroise permit and 1 in the new Loire Maritime permit which lies approximately 300 km west of Nantes. In 1980, SNEA (P) submitted plans for the required 50% relinquishment over their first three permit areas. These were approved by the French authorities in early 1981. Seismic work is continuing in the retained portions and this is likely to lead to further exploration drilling as the geological picture of the area is gradually pieced together.

Appendix 1

Basic Geological Concepts and Exploration/Production Techniques

The aims of this appendix are essentially threefold. Firstly, we briefly illustrate the main basic geological concepts which are directly relevant to hydrocarbon entrapment. Secondly, we look at the two principal aspects of hydrocarbon exploration, viz. seismic methods and exploration drilling. Thirdly, we aim to examine offshore production technology and look at some of the currently available systems and the most likely concepts for production systems applicable to deep water exploitation in the next decade or so.

Geological Control for Hydrocarbon Entrapment

The generation, migration and final entrapment of oil and gas represent a complex interplay of chemical and physical processes which take place, generally at a considerable depth beneath the earth's surface, over a period of millions of years of geological time. Throughout the world there are numerous instances known where hydrocarbons have been generated but have failed to be trapped. Equally common are occasions where excellent reservoirs and traps exist but where hydrocarbons have failed to be generated. It is therefore obvious that a certain number of criteria must be fulfilled in order for a hydrocarbon accumulation to take place. The four basic geological conditions necessary are as follows:

1. The presence of an organic-rich *source rock* such as coal, or deltaic, marine or lagoonal fine-grained sediments from which the hydrocarbons can be generated. The type, quality and quantities of hydrocarbons are a function of the type of source rock. Coal generally produces gas, a deltaic mudstone, rich in land-derived woody organic material, generally produces a waxy crude oil while lagoonal or marine mudstones usually produce a wax-free crude oil. In order for the organic fraction to be 'cooked' and converted to hydrocarbons, the source rock must be buried sufficiently deeply, usually to a depth of at least two thousand metres.
2. A suitable *reservoir rock* into which the oil or gas can accumulate having migrated from the source rock. This is usually a sandstone but could, for instance, be a limestone. Its main attributes are that it is both porous and permeable, which allows hydrocarbons to migrate through minute and intricate pathways and accumulate in the pore space of the rock. It is normal for the reservoir rock to be adjacent to the source rock, although this is not a vital requirement as oil and gas can, given suitable pathways, migrate long distances of at least several tens of kilometres.
3. The presence of a *cap rock* or impermeable stratum, such as claystone, mudstone or shale, which acts as a seal and prevents the accumulating hydrocarbons from escaping to the surface.
4. A *trapping structure* by which hydrocar-

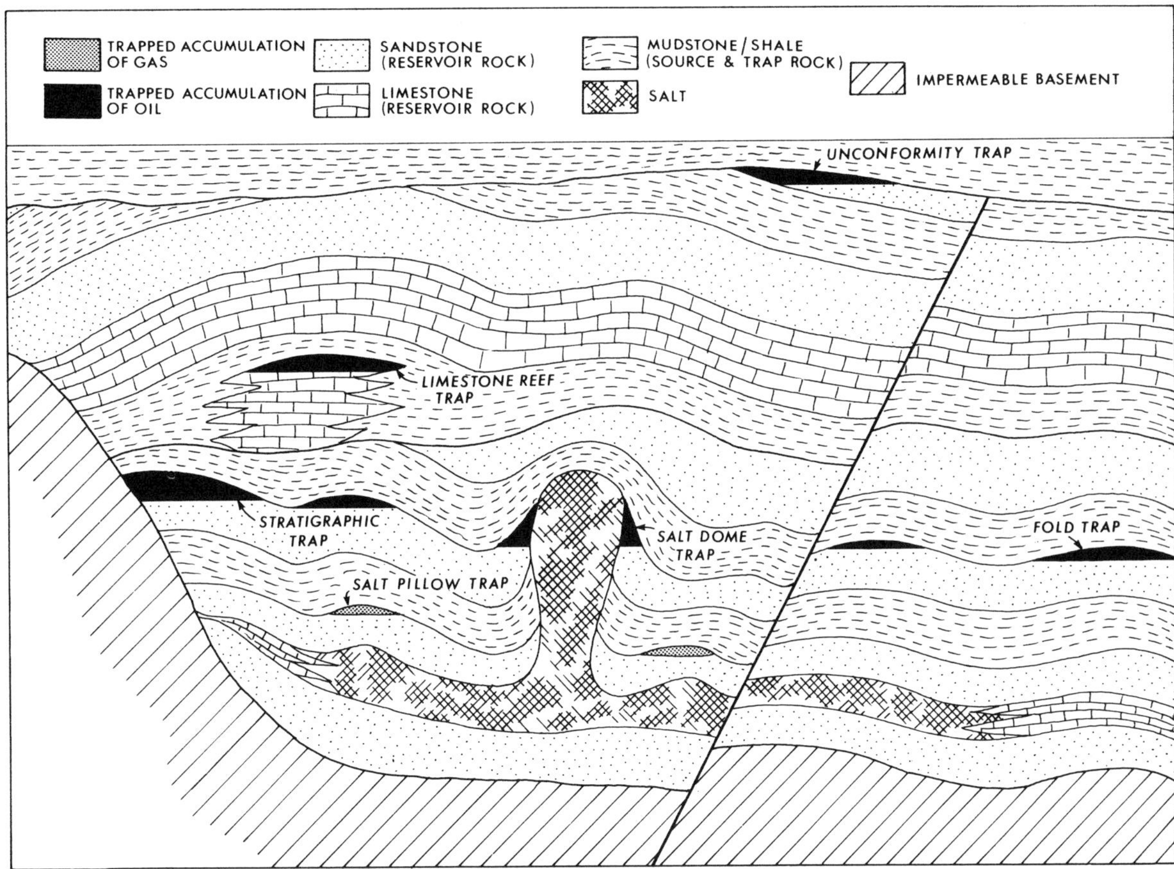

Figure A1.1. Sketch illustrating the various geological controls for hydrocarbon entrapment.

bons collect and are prevented from either lateral or vertical escape from the structure. Traps fall into two main categories, stratigraphic or structural, according to their mode of origin (Fig. A1.1).

Stratigraphic traps are formed during deposition of the sediments. Such traps may occur for instance where (a) sediments are laid down across an uneven erosional surface, (b) against an ancient shoreline, (c) in a reef environment where reservoir quality limestones are surrounded by impermeable source and capping sediments.

Structural traps are formed during post-depositional times by both large-scale and local deformation. They include (a) unconformity traps where impermeable strata rest unconformably on a reservoir sequence, (b) fold traps with hydrocarbons being trapped in the anticlinal arches, (c) fault traps where an impermeable rock abuts against the reservoir and where no escape is possible along the fault or (d) salt traps where a gravity difference between a thick salt sequence and a higher density sedimentary sequence causes the salt, an impermeable material, to rise into a dome which pierces through the overlying sediments and in doing so forms generally small hydrocarbon traps.

Seismic Exploration Methods

Seismic exploration provides a method of observing the thickness and structure of the individual rock strata forming the earth's crust. This method is based on the different reflecting and absorbing properties of the various rock types with respect to a shock wave generated at or near the earth's surface. A shock wave generated from a point source passes downwards into the underlying sediments, continuing at roughly the same velocity within any one sediment type until it encounters an interface, such as the boundary between two differing rock types. At this boundary part of the shock wave is reflected back to the surface where it is picked up and recorded by a system of geophones (signal receivers), and part is refracted down into the underlying strata. Of this refracted part, a similar process of wave separation occurs at the next boundary down, with a somewhat weaker reflected part being returned to the surface. The speed at which shock waves travel through any rock is dependent upon the elasticity of that rock; this tends to be low in loose and uncompacted sediments, increasing with depth as the sediments become harder and more compacted, and higher in both igneous rocks such as basalts and granites, and metamorphosed rocks such as schists or slates. The passage of a shock wave through rock follows the normal physical laws of reflection and refraction, and the success of producing a good seismic picture is depending upon an increase in the speed at which waves will pass through rock, with increased depth. With a prior knowledge, from refraction data, of the velocity of specific rock formations, the measurement of both the time taken for the wave to be reflected back to the surface and their strength, can be used to build up a seismic picture from which the depth and nature of the rock layers can be interpreted. Good quality data may yield information concerning, for instance, lateral facies variation within individual horizons and thereby allow the building up of an accurate seismic stratigraphy of a sedimentary basin and its individual geologically-promising potential hydrocarbon prospects.

Seismic exploration methods can be adapted for use both on land and offshore. Offshore operations, which are of chief concern to the oil industry in Britain and Ireland, are carried out from a boat equipped with an energy source for the generation of shock waves, and towing a string of geophones suspended on floats or strung out on a buoyant cable up to 3,000 m long. This cable is floated some 10 m or so beneath the water surface in order to minimize the interference of unwanted signals such as surface wave or boat noise. Explosives are now rarely used as shock wave generators, and the energy source now adopted for deep water operations fall into two categories; (a) the air gun source, used where deep penetration of the underlyine strata is required, which generates a seismic pulse by suddenly releasing highly compressed air into the water, and (b) sparker and boomer sources, generated by an electrical-arc discharge device and an electro-mechanical device respectively, are used only where shallow rock penetration is required. All signals from the reflected waves are measured as a time period in seconds between the signal being released, hitting the reflective boundary and being returned to the surface to be picked up by the geophones (two-way time). The information is recorded on magnetic tapes which are then processed to eliminate extraneous noise, spurious reflectors and multiple effects where ghost or apparent reflectors caused by particularly strong shallower reflecting horizons or frequently by the sea bottom appear on the seismic section. The final processed results give a geophysical cross-section, calibrated vertically in seconds, based on the two-way travel time of the energy waves through the underlyine rock strata. By shooting a very close-spaced grid of seismic lines, followed by the application of sophisticated processing techniques, three dimensional seismic sections may be constructed. Such three dimensional surveys are extremely useful in elucidating complex structures but, because they are very costly in comparison with the normal two dimensional seismic surveys, tend to be only shot during the appraisal phase following a hydrocarbon discovery.

Drilling a Well

Although exploration surveys can provide a good indication of the most likely location of a potentially hydrocarbon-bearing structure, the only direct means of testing such prospects is by drilling an exploration well. Offshore, and in particular around the shores of Ireland and West Britain, there is a considerable variation in the depth to the continental shelf. This has led to the use of the three main types of drilling rigs illustrated in Fig. A1.2 with each being suitable for working under different water depths and drilling conditions. *Jack-up rigs* are used in shallow water areas, where the water depth does not exceed 120 m and the sea-bed is firm. This type of rig is floated into position and the legs lowered to stand firmly on the sea floor, with the drilling platform being jacked clear of highwater level. Where water depths are deeper, or the sea-bed is unstable, a mobile floating type of drilling unit, the *semi-submersible rig*, is employed. This is usually capable of working in water depths of up to about 500 m and comprises a drilling

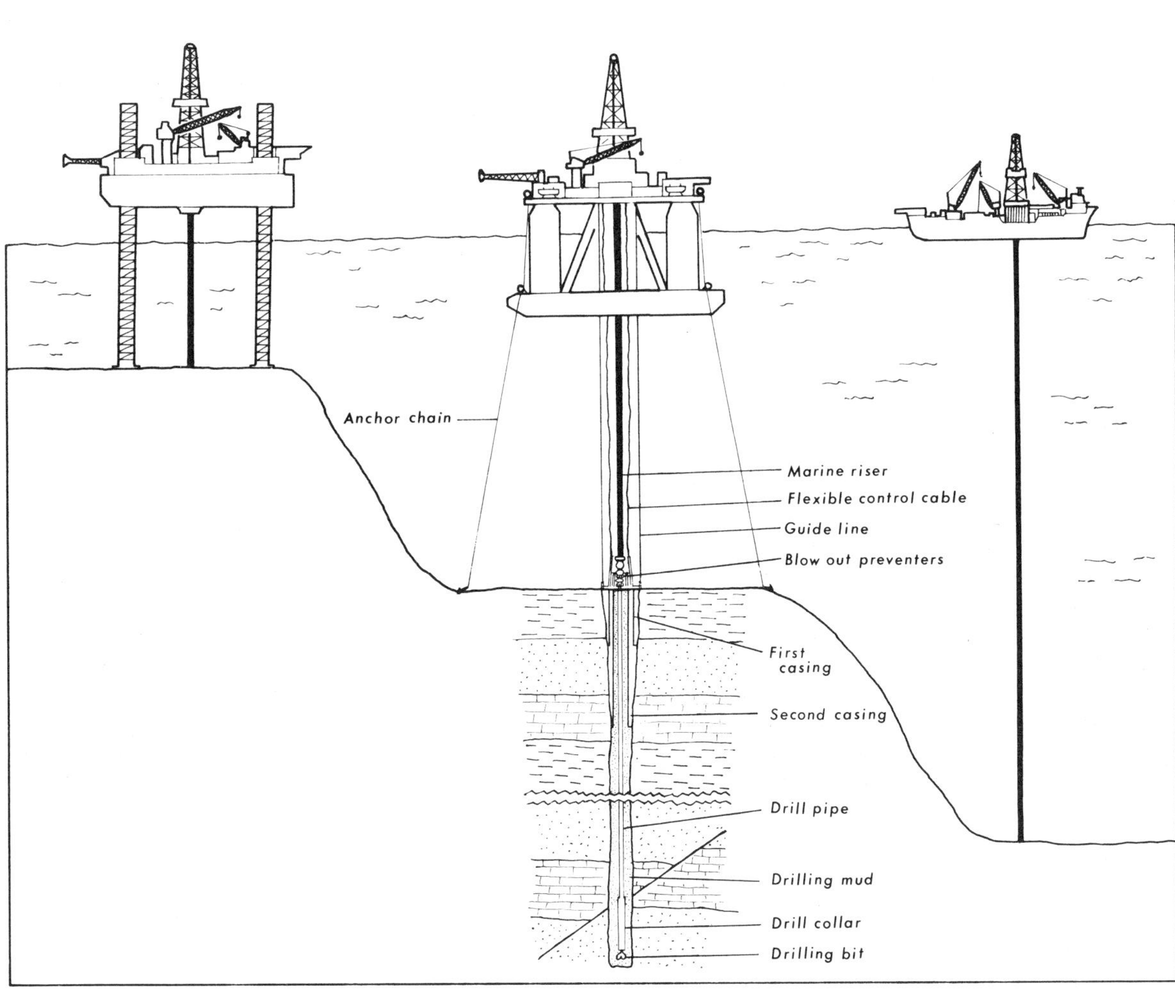

Figure A1.2. Diagram illustrating the three main types of offshore drilling units, together with a section through a typical exploratory well.

platform supported on a partially submerged buoyant framework. The latter normally comprises a number of large hulls which during drilling are partially filled with water so as to float about 20 m below the water surface and below the normal wave base. The design of the semi-submersible therefore provides a relatively stable platform under rough open marine conditions which are subject to strong wave motions. These rigs are normally held in position at the well location by a series of 8 to 12 anchors but more recent designs have a system of computer-assisted dynamic positioning to keep the rig on station. This effectively eliminates the need for anchors and allows these large semi-submersible rigs to drill in waters in excess of 500 m. *Drill ships* are capable of operating in all except extremely shallow water areas and are particularly effective where water depths exceed the range of the normal semi-submersibles. However, being totally surface floating structures they are affected to a greater extent by heavy seas than the large semi-submersible rigs.

The location of an exploratory well is sited to intersect various prospective features identified on the initial geophysical surveys. The well is commenced or 'spudded' by drilling a hole of up to 36 inches in diameter and 50 to 100 m deep. The hole is drilled using a rotating steel-toothed or diamond-studded *bit* attached to the lower end of *the drill string* (Fig. A1.2) which is a set of steel pipes. This hole is then prevented from collapsing or caving in by cementing a steel pipe or *casing* to the rock wall of the hole. A smaller diameter hole is then drilled from the base of the casing and this in turn is cased from the bottom of the well as far as the surface by a smaller diameter casing. This casing is cemented to the rock well in the lower part of the well and to the inside of the large casing in the upper part of the hole. The process is continued downwards with progressively smaller diameters of hole and of casing. Before the casing is lowered and cemented into position a series of instruments, the *electric logging tools*, are temporarily lowered into the hole in order to establish the various physical properties of the rocks which have been drilled.

During the drilling of a well, a high density mixture of chemicals known as the *drilling mud* is continually circulated through the hole, being pumped down inside the drill string, out through the drilling bit and back up the hole to the mud pits on board the drilling rig. This mud acts partly as a lubricant, a coolant, and to transport the newly excavated fragments or chippings from the bottom of the hole to the surface, and partly to prevent any inward collapse of the uncased portion of the well. Connected to the surface of the well, and usually sitting on the sea-bed itself, the *blow-out preventer* stack consists of a complex series of pipes and valves to prevent any sudden rush of fluid or gas onto the sea-floor in the event of encountering a reservoir at unexpectedly high pressures.

Hydrocarbon Production Technology

When an oil or gas discovery has been made and the field has been adequately delineated, appraised and declared commercial, the development phase of operations commences to enable the hydrocarbons to be extracted most efficiently and to be brought ashore. In shallow waters, i.e. normally less than about 100 m, this usually involves the installation of one or more production platforms, each with a number of deviated or divergent wells, some for producing the hydrocarbons and others for injecting liquid or gases to maximize the extraction rate from the reservoir, together with a pipeline system to bring the oil or gas ashore. In such cases the production platforms are normally simple steel jacket or concrete structures (Fig. A1.3A). However, such systems become prohibitively expensive in deeper waters due mainly to size and weight and installation difficulties (although steel jacket structures have been used in almost 400 m of tranquil water in the Gulf of Mexico). As many, if not most, of the future oil and gas discoveries are likely to be made in waters deeper than 100 m, other more advanced production systems are certain to be required. Quite a number of production concepts have recently been developed but most of these are still at the design stage. Although this section is not aimed at giving a comprehensive resumé of all such concepts, the main ones may conveniently be divided into the following categories:

(1) Extension of *fixed steel and concrete platforms*, e.g. *Condeep Structure* installed in Statfjord field of the North Sea (Fig. A1.3B). These are extremely expensive and can realistically only be considered for large fields.

(2) *Floating systems*, which can be either very large or small. The large ones, such as the *Caisson Vessel* system would have the capacity to store large quantities of hydrocarbons in hostile and deep water environments until they could be offloaded into tankers during less inclement weather (Fig. A1.3C). Once again a large field is necessary to economically support such a system while major problems concerning the risers and the mooring of the vessels still remain to be overcome. This system is still at the design stage. The small floating system could be either *semi-submersible*-type vessels as in the Buchan Field of the North Sea or a converted tanker-type vessel as visualized by BP's *Single Well Oil Production System* (SWOPS) (Fig. A1.3D). In the first case (semi-submersible) the oil or gas is offloaded onto a fleet of tankers while in the SWOPS system the converted tanker unlatches from the well, retrieves the flexible riser and transports the produced hydrocarbons to shore. However in general those small systems will probably be used to exploit small fields, or as an early and rapid small scale production system with no facilities for secondary or tertiary recovery methods.

(3) *Floating systems with tension legs*. These comprise a large semi-submersible type structure anchored in place by a series of usually vertical tensional steel pipes or cables (Fig. A1.3G). A *Tension Leg Production* system is being constructed

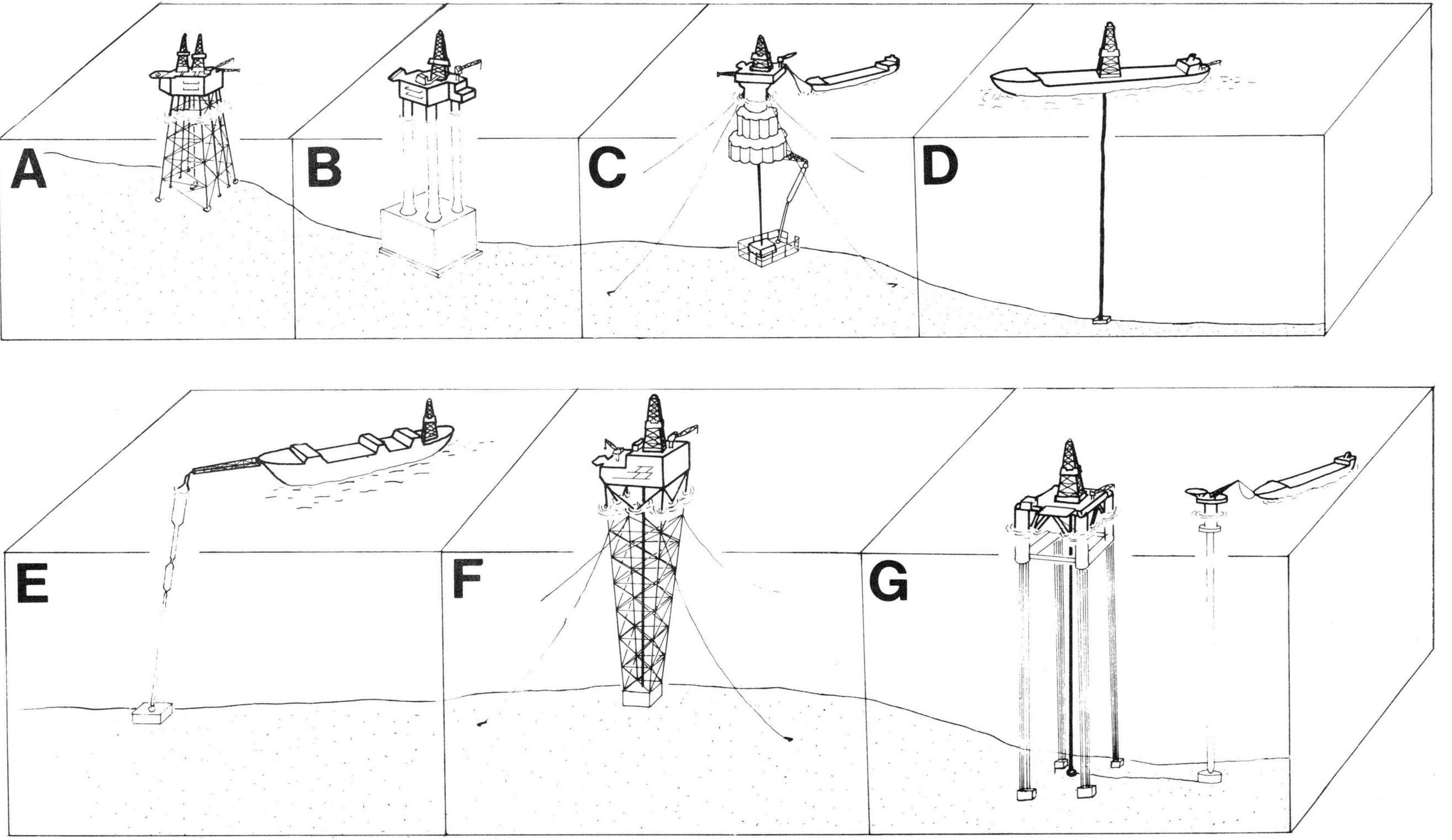

Figure A1.3. Diagram illustrating various current and likely production systems.
A. Steel Jacket Platform. B. Concrete Platform. C. Caisson Vessel Platform. D. Single Well Oil Production System. E. Single Anchor Leg Mooring System. F. Guyed Tower Platform. G. Tension Leg Platform.

for installation in the Hutton Field of the North Sea. As with most of the deep water floating systems, problems with extreme motions and stresses on the riser system could be anticipated.

(4) *Buoyant articulated columns* connected to the sea bed where the oil or gas is offloaded to shuttle tankers. An example of this is the *Single Anchor Leg Mooring* (SALM) unit installed in the Thistle field and planned for the Fulmar Field of the North Sea (Fig. A1.3E). However, problems with the various articulation joints still remain to be overcome.

(5) *Guyed Tower system.* This comprises a downward-narrowing steel lattice tower system weighted at the base by a large spud can full of liquid ballast and held in place by up to 16 catenary guy lines (Fig. A1.3F). Although the system has not yet been installed on a producing field, scale models have been successfully tested.

(6) *Totally subsea systems.* These require that all well completion and export facilities be carried out on the seabed in a series of module units. These include a one-atmosphere habitat module for housing the operations personnel. These systems are only at the earliest stages of design.

Appendix 2

Glossary

The science of geology, and especially petroleum geology, abounds in specialist technical terminology and jargon which may prove extremely baffling to a newcomer to the oil industry. This final appendix is aimed at providing a basic understanding of the most commonly used terms. However, the following short glossary does not pretend to be comprehensive. The entries should be taken by the reader as giving a guide to the meanings of the geological and petroleum terms and not as formal definitions.

Acidization: Treatment of a reservoir formation with acid to assist the flow of hydrocarbons by improving the permeability of the rock.

Annulus: The annular space between the drill string and the well bore.

Anticline: A tectonic structure in which strata are folded so as to form a dome or an arch.

API gravity: The universally accepted arbitrary scale adopted by the American Petroleum Institute for defining the specific gravity of oils:

$$\text{API gravity (in degrees)} = \frac{141.5}{\text{specific gravity at } 60°\text{F}} - 131.5$$

Appraisal wells: Wells drilled to determine the physical extent, reserves and likely production rate of a potential field.

Argillaceous rocks: Fine-grained sedimentary rocks comprising claystones, shales, mudstones, siltstones and marls.

Barrel: Unit of volume measurement of petroleum and its products:
1 barrel = 42 U.S. gallons
= 35 imperial gallons (approx.)
= 159 litres (approx.)

Basalt: A fine-grained basic (i.e. silica-poor) igneous rock. It represents rapidly cooled molten magma.

Bed: Geological term describing a stratum (layer of sediments or sedimentary rock) of uniform composition and texture.

Blow out preventer: Hydraulically-operated wellhead valves designed to prevent a blowout (an uncontrolled escape of gas, oil or salt water).

Bottom hole assembly (BHA): The lower end of the drill string comprising the drill bit, drill collars, heavyweight drill pipe and ancillary equipment.

Cap rock: An impervious layer of rock (e.g. shale) which directly overlies a reservoir rock thereby preventing the escape of hydrocarbons.

Casing: Steel pipe that is cemented to the sides of a well to prevent the walls from collapsing, to exclude unwanted fluids and to enable the control of well pressures and produced hydrocarbons.

Chalk: Very fine-grained, pure white limestone.

Christmas tree: Assembly of valves and pipes fitted to a production wellhead to control the flow of oil or gas and to prevent a possible blowout.

Commercial field: An oil and/or gas field judged to be capable of producing enough net income to make it financially worth developing.

Condensate: Liquid hydrocarbons which are sometimes produced together with natural gas.

Conglomerate: Sedimentary accumulation of rounded or semi-rounded fairly large fragments of rock.

Continental shelf: That part of the sea-floor adjoining a land mass and where the water depth is generally less than 200 m. It is considered to be a submerged portion of the continental mass, the outer margin of which is regarded as the boundary between continental crust and oceanic crust.

Core: The cylindrical section of rock obtained by a core barrel fitted with an annular bit.

Crude oil: The oil that is produced from a reservoir (after any associated gas has been removed).

Derrick: The latticed steel pyramid used to support equipment which has to be raised or lowered during drilling and well-servicing operations.

Development well: A well drilled with a view to producing hydrocarbons from a proven field.

Deviated well: A well drilled in such a way that its controlled direction departs progressively from the vertical. Such wells are frequently drilled where a number of wells are required to reach different parts of a reservoir from a single platform.

Dip: A measure of the inclination of rock strata with respect to the horizontal.

Drill bit: The part of the drilling tool which cuts through the rock.

Drill collars: Lengths of extra-heavy steel tubing located immediately above the drill bit in order to maintain pressure on the bit and to keep the drill string in tension.

Drill Stem Test: A test whereby formation fluids are allowed to flow into the drill pipe and to the surface for a relatively short period (normally up to 24 hours).

Drill string: Lengths of steel tubing approximately 10 m long screwed together to form a pipe connecting the drill bit to the drilling rig.

Drilling mud: A mixture of clays, water and chemicals pumped down the drill pipe and up the annulus during drilling.

Electric logging: The method of surveying a well by means of running special tools on an electric cable. A variety of formation properties are recorded on tape and details of, for instance, rock types, fluid content, porosity, and dip may be extracted from the results.

Enhanced oil recovery: A process whereby oil is recovered other than by the natural pressure in a reservoir.

Evaporite: The remains of a solution after most of the solvent (usually water) has evaporated, e.g. salt.

Facies: The collective features which characterize a sediment deposited in a particular sedimentary environment.

Fault: A fracture in the earth's crust along which the rocks on one side are displaced relative to those on the other.

Field: A geographic area under which an oil and/or gas reservoir lies.

Formation: A homogeneous body of rock.

Fracturing: A process under which chemicals are forced, under high pressure, into a rock formation in order to create a series of interconnected cracks or fractures in the rock thereby increasing the permeability and improving the flow rate of hydrocarbons. (Known also as 'fraccing'.)

Gas cap: A layer of natural gas above the oil in an oil reservoir.

Hydrocarbons: Compounds consisting entirely of the elements hydrogen and carbon, which form the bulk of oil and natural gas.

Igneous rocks: Rocks which were formed by the solidification of molten magma.

Injection well: A well used for injecting fluid (usually water or gas) into the reservoir rock in order to maintain reservoir pressure in secondary recovery.

Interstitial water: Water present in the pores of the oil- or gas-bearing zone of a reservoir.

Kick: The situation that occurs when the pressure of the formation fluid exceeds the hydrostatic head of the mud column, thus allowing formation fluid to enter the well bore. If not controlled immediately it may lead to a blowout.

Killing a well: Overcoming the tendency of a well to flow by filling the well bore with drilling mud of suitable density.

Limestone: A sedimentary rock consisting essentially of organically or chemically derived carbonate minerals.

Marginal field: A field that may or may not produce sufficient net income to make it worth developing at a given time; should technical or economical conditions change, such a field may subsequently become commercial.

Marine riser: The pipe connecting the sub-sea wellhead, which is situated on the sea-bed, to the drilling rig.

Marl: A calcareous mudstone.

Metamorphic rocks: Rocks which have been altered from their original composition and structure by heat and pressure.

Methane: The chief constituent of natural gas. It is the hydrocarbon with the lightest molecule (CH_4).

Oil: A mixture of liquid hydrocarbons of different molecular weights.

Palaeogeography: A reconstruction of the relative positions of land and water at a particular time in geological history.

Pay zone: The stratum of rock in which hydrocarbons are found.

Permeability: The ability of a rock stratum to allow fluids to pass through it.

Porosity: The occurrence of voids or cavities in a rock, which gives it the ability to retain fluids or gases.

Primary recovery: Recovery of hydrocarbons from a reservoir using only the natural pressure in the reservoir to force the hydrocarbons out.

Recoverable reserves: The proportion of the hydrocarbons in a reservoir that can be removed using currently available techniques.

Reservoir: A stratum in which hydrocarbons are present.

Salt dome: A dome of salt which has risen and broken through overlying formations.

Salt pillow: A mass of salt which has risen up causing overlying formations to bulge upwards, but which has not actually broken through.

Sandstone: A sedimentary rock composed primarily of quartz grains.

Secondary recovery: Recovery of hydrocarbons by artificially maintaining or enhancing the reservoir pressure by injecting water, gas or other substances into the reservoir.

Sedimentary rocks: Rocks formed from the products of weathering and erosion of older rocks or by the accumulation of mud, calcareous deposits of organic origin or chemical deposits.

Seismic survey: An exploration technique for determining the detailed structure of the rocks underlying an area by passing acoustic shock waves into the strata and measuring the reflected signals.

Show: An indication of oil or gas from an exploration well.

Sour crude: Crude oil with a high sulphur content.

Spudding in: The process of starting to drill a well by making a hole in the sea bed using a large-diameter drill bit.

Step-out well: A well drilled beyond the proven limits of a field to investigate a possible extension of the field.

Sub-sea completion: The process of installing a wellhead on the ocean floor.

Sweet crude: Crude oil with a low sulphur content.

Syncline: A tectonically formed structure in which the strata are folded so as to form a trough or bowl.

Tectonic process: A process whereby rocks are deformed by natural forces within the earth's crust.

Tertiary recovery: Recovery of hydrocarbons by methods other than primary and secondary recovery. It generally involves sophisticated techniques such as heating the reservoir or injecting chemicals to reduce the viscosity of the oil.

Trap: A geological structure in which hydrocarbons accumulate to form an oil and/or gas field.

Turbine drilling: Drilling in which the drill string remains stationary, the bit being rotated by a down-hole turbine powered by the jetting action of the drilling mud.

Unconformity: An unconformity represents a significant time-break between two sets of beds. It is often reflected as a discordant relationship between the older and more deformed beds and the overlying, younger and less deformed rocks.

Wellhead: The control equipment fitted to the top of the well casing, at the sea-bed, incorporating outlets, valves, blow out preventers etc.

Wildcat well: An exploration well drilled without much knowledge of the underlying rocks and away from known oil or gas fields.